Numerical Simulation in Science and Engineering

Edited by
Michael Griebel and
Christoph Zenger

Notes on Numerical Fluid Mechanics (NNFM) Volume 48

Volume 26 Numerical Solution of Compressible Euler Flows (A. Dervieux / B. van Leer / J. Periaux / A. Rizzi, Eds.)
Volume 27 Numerical Simulation of Oscillatory Convection in Low-Pr Fluids (B. Roux, Ed.)
Volume 28 Vortical Solution of the Conical Euler Equations (K. G. Powell)
Volume 29 Proceedings of the Eighth GAMM-Conference on Numerical Methods in Fluid Mechanics (P. Wesseling, Ed.)
Volume 30 Numerical Treatment of the Navier-Stokes Equations (W. Hackbusch / R. Rannacher, Eds.)
Volume 31 Parallel Algorithms for Partial Differential Equations (W. Hackbusch, Ed.)
Volume 32 Adaptive Finite Element Solution Algorithm for the Euler Equations (R. A. Shapiro)
Volume 33 Numerical Techniques for Boundary Element Methods (W. Hackbusch, Ed.)
Volume 34 Numerical Solutions of the Euler Equations for Steady Flow Problems (A. Eberle / A. Rizzi / E. H. Hirschel)
Volume 35 Proceedings of the Ninth GAMM-Conference on Numerical Methods in Fluid Mechanics (J. B. Vos / A. Rizzi / I. L. Ryhming, Eds.)
Volume 36 Numerical Simulation of 3-D Incompressible Unsteady Viscous Laminar Flows (M. Deville / T.-H. Lê / Y. Morchoisne, Eds.)
Volume 37 Supercomputers and Their Performance in Computational Fluid Mechanics (K. Fujii, Ed.)
Volume 38 Flow Simulation on High-Performance Computers I (E. H. Hirschel, Ed.)
Volume 39 3-D Computation of Incompressible Internal Flows (G. Sottas / I. L. Ryhming, Eds.)
Volume 40 Physics of Separated Flow – Numerical, Experimental, and Theoretical Aspects (K. Gersten, Ed.)
Volume 41 Incomplete Decompositions (ILU) – Algorithms, Theory and Applications (W. Hackbusch / G. Wittum, Eds.)
Volume 42 EUROVAL – A European Initiative on Validation of CFD Codes (W. Haase / F. Brandsma / E. Elsholz / M. Leschziner / D. Schwamborn, Eds.)
Volume 43 Nonlinear Hyperbolic Problems: Theoretical, Applied, and Computational Aspects Proceedings of the Fourth International Conference on Hyperbolic Problems, Taormina, Italy, April 3 to 8, 1992 (A. Donato / F. Oliveri, Eds.)
Volume 44 Multiblock Grid Generation – Results of the EC/BRITE-EURAM Project EUROMESH, 1990–1992 (N. P. Weatherill / M. J. Marchant / D. A. King, Eds.)
Volume 45 Numerical Methods for Advection – Diffusion Problems (C. B. Vreugdenhil / B. Koren, Eds.)
Volume 46 Adaptive Methods – Algorithms, Theory and Applications. Proceedings of the Ninth GAMM-Seminar, Kiel, January 22–24, 1993 (W. Hackbusch / G. Wittum, Eds.)
Volume 47 Numerical Methods for the Navier-Stokes Equations (F.-K. Hebeker, R. Rannacher, G. Wittum, Eds.)
Volume 48 Numerical Simulation in Science and Engineering (M. Griebel, C. Zenger, Eds.)

Volumes 1 to 25 are out of print.
The addresses of the Editors are listed at the end of the book.

Numerical Simulation in Science and Engineering

Proceedings of the FORTWIHR Symposium on High Performance Scientific Computing, München, June 17–18, 1993

Edited by
Michael Griebel and
Christoph Zenger

Softcover reprint of the hardcover 1st edition 1994

Vieweg ist a subsidiary company of the Bertelsmann Publishing Group International.

Produced by W. Langelüddecke, Braunschweig
Printed on acid-free paper

ISSN 0179-9614
ISBN-13: 978-3-528-07648-1 e-ISBN-13: 978-3-322-89727-5
DOI: 10.1007/978-3-322-89727-5

PREFACE

Founded in April, 1992 and financed by the State of Bavaria and the Bavarian Research Foundation, the Bavarian Consortium for High Performance Scientific Computing (FORTWIHR) consists of more than 40 scientists working in the fields of engineering sciences, applied mathematics, and computer science at the Technische Universität München and at the Friedrich-Alexander-Universität Erlangen-Nürnberg. Its interdisciplinary concept is based on the recognition that the increasing significance of the yet young discipline High Performance Scientific Computing (HPSC) can only be given due consideration if the technical knowledge of the engineer, the numerical methods of the mathematician, and the computers and up to date methods of computer science are all applied equally.

Besides the aim to introduce HPSC into the graduate degree program at the universities, there is a strong emphasis on cooperation with industry in all areas of research. Direct cooperation and a transfer of knowledge through training courses and conferences take place in order to ensure the rapid utilization of all results of research. In this spirit, FORTWIHR annually organizes symposiums on High Performance Scientific Computing and Numerical Simulation in Science and Engineering.

This book contains 14 contributions, presented at FORTWIHR's first symposium on June 17-18, 1993 in München. The meeting was attended by more than 300 scientists from Germany as well as from neighbouring countries. The contributions of this volume give a survey on recent research results and industrial applications of numerical simulation in the areas of fluid mechanics, dynamic systems in aerospace, melting processes and crystal growth, and semi-conductor and electric circuit technology.

We like to thank the BMW AG, München, which generously provided her facilities at the *Forschungs- und Ingenieur-Zentrum* (FIZ) in München. Furthermore, we want to mention the exhibitors and sponsors Convex Computer, Digital Equipment, GENIAS Software, Hewlett Packard, IBM, Siemens-Nixdorf, and Silicon Graphics, whose financial support was very important for the success of the symposium.

Finally, we are indebted to Hans Bungartz and the team of persons involved in the preparation and organization of the symposium. At last, we like to thank Stefan Zimmer for his assistance in compiling this volume.

München, May 1994

Michael Griebel
Christoph Zenger

CONTENTS

ON THE NECESSITY OF SUPPORTING RESEARCH

Nikolaus Fiebiger
Bayerische Forschungsstiftung
Kardinal-Döpfner-Str. 4/I
80333 München

Originally, the title of this contribution was *On the Necessity of Governmental Support of Research.* After thinking about it, it became apparent that the topic must encompass more. Surely, financial support of research is necessary for our country. But it by no means should be restricted to governmental support. Support of research must also include the private sector, for which the USA is an example. There, support of research by private foundations has decisively contributed to the success of North-American science. Therefore, I generalized the title of my contribution to *On the Necessity of Supporting Research.*

In the following I try to answer why support of research is reasonable and necessary and who is interested in research at all. Then, I want to enumerate different forms of research that need distinctive support and I want to ask weather the structure of research is adequate for its tasks. Finally, I address the problem how to organize research more efficiently.

WHY SUPPORT RESEARCH AT ALL?

In Germany, we enjoy one of the highest standards of living in the world. However, we are surpassed by far by a few small emirates in the Persian Golf. There, the foundation of wealth has been provided by nature in the form of a rich supply of oil. Exporting raw materials and resources is thus one way of making an economy flourish. The wealth in our country, however, relies on producing goods and exporting merchandise all throughout the world. This merchandise is wanted and bought only if it is, in the broadest sense of the word, better than the corresponding products of our competitors.

As we all know, the success of the German economy is based on the high level of education in our society and on the research and development in our scientific facilities and in our industry. We tend to speak of the raw material or resource 'mind', the only means available to us of achieving a high standard of living. Therefore, its development, its application and its exploitation is a necessity for the prosperity of our country.

WHO IS INTERESTED IN RESEARCH?

Research costs money. Ultimately someone, be it a person or an institution, will invest in research only if there is a prospect of getting back the principle plus interest plus compound interest. This expectation is shared by the scientists themselves, what may here be neglected due to the limited scope of my topic, then by the beneficiaries of the results of research applied in the economy, and at last by the society, represented by the government and also by private sponsors.

A beneficiary of the results of research is quite often the industry. It, and this applies internationally, is increasingly concerned with research and development only in so far as they provide a means of developing competitive products within a relatively short time. Thus, basic research is hardly being pursued by industry. This basis for research in our country is established and maintained virtually exclusively by the state. It consists of education in general, of inducements for young scientists in particular and of the basic equipment of our universities. Moreover, this ensures research at our universities at all. Furthermore, the federal government (Bund) and the states (Länder) finance the Deutsche Forschungsgemeinschaft and the institutes of the so-called Blauen Liste. In addition, the federal government runs a number of Großforschungseinrichtungen and participates in a number of international research projects (i.e. CERN).

TYPES OF RESEARCH THAT REQUIRE A SPECIFIC WAY OF SUPPORT.

A classification of research, as I intend to give it here, can only be done very roughly. In keeping with the topic I am going to refer to the necessity of a specific type of support for certain fields of research. I'd like to distinguish between basic research, applied research that reaches to the development of products and basic research with a particular aim or object. The third term is not a conventional one and thus needs explanation: There exists often a very clearly defined research aim. As an example, I cite an inoculation against AIDS. The way to this aim passes through presently unknown areas that have to be investigated. This research belongs to a pre-competitive area; it can and should be jointly pursued by the institutions and universities expressly committed to this specific project together with industry and other scientific facilities.

Now, a classification of research can be done with regard to the factor of time needed for it. The various time demands on research can be represented in a diagram, see Figure 1. Here as abscissa the different types of research can be entered beginning with basic research, continuing to applied research and extending to development aimed at one product. As ordinate, the available time is entered. In the case of basic research it is usually quite large. The closer it comes to a product, the more decisive the competitive situation becomes and the less the spent time has to be. Under the abscissa of this diagram, we can associate the universities and the Max-Planck-Institute to basic research, and the Fraunhofer-Gesellschaft, the Großforschungseinrichtungen and the institutes of the Blaue Liste to applied research. Proceeding on to development aimed at one product we can include the industry.

For an optimal support of research, the process of decision should meet the time-competitive criteria. This is the main problem concerning governmental support. I maintain: the state can not meet the time demand; it can not act quickly. State institutions depend on the respective budget, which the state legislative must approve of. The budget is the prerogative of the legislature and being one of its most important and distinguished rights it is not likely to be changed. A comprehensive budget for the universities, for example, doesn't seem to be a feasible solution either, for reasons I can not go into now.

What solutions are then possible? First of all, state institutions have not got only state funds. They have other funds at their disposal that are not subject to budget regulations. These are mainly funds provided by foundations. Unfortunately, our country doesn't have the tradition of foundations, even though is very prosperous. We

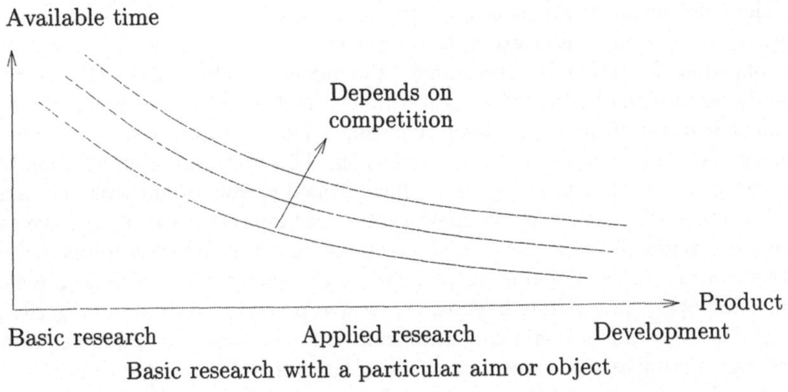

Fig. 1 Relation between the type of research and the available time

can learn from the United States how interested in art and science potential sponsors are. Foundations can only contribute part of the necessary means, but especially for time-competitive projects their contributions can be highly useful. Another possibility might be a combination of governmental and industrial support with the private sector providing support especially for time-competitive projects due to its relatively large flexibility.

IS THE STRUCTURE OF RESEARCH ADEQUATELY GEARED TO ITS TASKS?

Beside industry, the universities, the Max-Planck Gesellschaft, the Großforschungseinrichtungen, the institutes of the Blaue Liste and the Fraunhofer-Gesellschaft are involved in research. The sixteen states and the federal government, in some cases mutually in others separately, are responsible for the respective institutions. For a classification of the various research institutes with regard to the different types of research, see Figure 1 and Table 1.

Table 1 Classification of research institutions.

Basic research	Applied research	Development
Universities	Fraunhofer-Gesellschaft	
	Großforschungsinstitute	
Max-Planck-Gesellschaft	Fachhochschulen	Industry
	Blaue Liste	

The relationship of the federal government (Bund) to the individual states (Länder), as prescribed by our constitution, is a problem that prevents our country from optimizing conditions for research. Accordingly, the universities are state institutions and can't directly be accessed by the federal government. But, on the other hand, the federal government has commitments in areas requiring scientific support such as in securing the sources of energy, public health care and so on. This question of jurisdiction has created a lot of problems that have led to an uncontrolled sprawl of our research landscape.

The universities must be the main centers of basic research. They have certain definite advantages: Universities are, as the name indicates, interdisciplinary. Universities have a continual flow of young people, and by the qualification of those a mean of selection. Moreover, universities have the exclusive privilege of determining the standards for graduation and of conferring academic titles (Diplom, Promotion, Habilitation). Furthermore, universities still have a high degree of flexibility due to a large number of untenured positions. In other state institutions including research institutions the existing labour laws tend to make employees dependent on lifetime positions. This means it is hardly possible to replace someone who is good by someone who is even better.

As a result of these considerations I therefore demand that research that can be done at universities, (which certainly does not apply to all types of research,) be done there.

If, like in the above mentioned case of AIDS, object-oriented research is necessary, what are then the possibilities for the federal government to meet its responsibilities? Surely it is impossible to avoid the installation of federal research institutes. Since, however, those problems tend to have a time limit, i.e. the work is or should be accomplished within a certain time, consequently, only temporary institutes should be created, in other words, institutes that can be totally dissolved. This demand defies German tradition and German labour laws. Let us consider the example of AIDS again: an institute for AIDS research means establishing stationary buildings, means creating lifetime positions and so on. After the problem has been solved, in this case the basis for an inoculation serum is found, the only thing that can be done according to current procedure is to redirect the work and the staff with lifetime positions to a new aim.

From the information received from the Ministry for Research and Technology it is evident that the share of funds allocated to the non-university research facilities is increasing at a faster rate than of that allocated to the universities. However, we need Großforschungseinrichtungen only for that tasks that should for practical reasons not be pursued by the universities, like for example large particle accelerators and so on. So as not to be misunderstood at this point, I'd like to explain that now I am only talking about the organization of research and not about the quality of research that is also being achieved very well in non-university research facilities.

These considerations compel me to demand of our political leaders that a long-term programme (25 years) be set up to dissolve those institutions whose tasks have been completed in the meantime. This proposal is certainly not easy to realize. But all the more important it seems to me to create as a matter of principle those institutions with a specific research aim in future only for a limited time.

CAN RESEARCH BE ORGANIZED MORE EFFICIENTLY?

My answer is yes and also applies to the existing organization of research, whereby I am referring to object-oriented and applied research. In this our country we must find ways of pre-competitive cooperation, especially for the solution of time-competitive problems. The slogan for this must be: coordinate and strengthen. By this I mean that cooperation among specialized institutions working on the same project must be coordinated and financial support to this area increased. In the case of AIDS, our large chemical and pharmaceutical enterprises, some of the Max-Planck-institutes and some departments at universities would be included in the efforts to achieve a high degree of coordination. Coordination measures would be carried out by a temporary institution as previously described. This, however, would not entail a comprehensive relocation of all the participants in one place. Pre-competitive cooperation would enable a head start in international competition.

There are problems that have to be solved in order to bring about pre-competitive cooperation in a temporary institute. Who is to assume the costs and responsibility for this institution? The effort to found a registered cooperation/society or a non-profit organization and to secure the financing should not have to be undertaken each time the need for such a temporary institute arises. Who will manage the institute, provide for money, rooms, equipment and personnel? By no means should the scientists who are expected to produce a solution to a problem be let alone with the problems of management. Who will make space available, since the construction of new buildings designated for a specific purpose is to be avoided?

In answering these questions industry should take the initiative where research is product-oriented; the federal government should take the initiative in matters concerning basic research. The pre-competitive cooperation between the scientific community in the universities and industry in temporary institutes is, by the way, the best instrument for transferring knowledge from the laboratory to the economy that one can imagine, for, in the end, this transfer always takes place successfully only from person to person. An example for such an organization is definitely the Bavarian Research Consortium for High Performance Scientific Computing (FORTWIHR).

CONCLUDING REMARKS

Research as regarded in this presentation is not limited to the natural sciences, to technology and economics. Germany is a nation with culture, whose roots nourish the natural sciences and technology, even in those cases where the connection is not always obvious. Therefore, research in the humanities must also be supported. This demand can easily be justified by economic considerations. We need cultural studies, understanding for foreign cultures, languages, knowledge of the various areas of legal issues, knowledge of the world's religions and so on, in order to gain access to foreign markets and in order to promote trade. We need guidelines on how to co-exist at all.

I'm going to summarize my contribution with the following statement: Financial support of research is necessary because research and development in industry need a firm foundation to build up on. In consideration of German circumstances, research must be financed by the government, at least in main parts. A sufficient financial amount must be provided that will enable German science to keep up with its competitors. Both, the federal government (Bund) and the states (Länder) as providers have the right and duty to insure efficient structures and organization. The federal government should reduce its involvement as far as possible by initiating and establishing support foundations. By these means it can encourage a hitherto unknown degree of flexibility. To promote economic development and our competitiveness on the international market the state must give more consideration than it has done so far to the demands of time-competitive projects.

PARALLEL COMPUTER ARCHITECTURES FOR NUMERICAL SIMULATION

A. Bode
Institut für Informatik,
Lehrstuhl für Rechnertechnik und Rechnerorganisation
Technische Universität München,
Arcisstr. 21, D–80333 München
Tel.: +49+89-2105-8240, Fax: +49+89-2105-8232,
email: bode@informatik.tu-muenchen.de

SUMMARY

Numerical simulation requires very high performance computer architectures. This can only be achieved by using various forms of parallelism. The second section of this paper gives an introduction to the state-of-the-art in parallel computer architectures and their programming. Today, some open problems remain for the user of such architectures. Those problems and possible solutions are dealt with in the third section. Problems of communication, portability, and flexibility are addressed.

MOTIVATION

"High Performance Computing and Computer Communications Networks are becoming increasingly important to scientific advancement, economic competition, and national security. The technology is reaching the point of having a transforming effect on our society, industries, and educational institutions." This statement is taken from the executive summary of the US research and development program *Grand Challenges: High Performance Computing and Communication* (HPCC, Bromley, 1992). As described in this program as well as in the Japanese *Real World Computing Program* (RWC) and the European *High Performance Computing and Networking Report* (Rubbia-Report), high performance computing in the future will be a synonym to the use of parallel computer techniques. Sequential computers cannot deliver enough computing power for the solution of demanding problems such as numerical simulation. As with any new discipline, a large variety of computer architectures regarding parallelism in processing, communication, and storage as well as a large variety of programming models and control strategies have been proposed. Standards have not yet been established as for more conventional computation models. The next section gives a summary on parallel computer architectures and their programming, the third section discusses solutions to some problems that have hindered the introduction of parallel computer architectures in industrial environments until now.

STATE-OF-THE-ART: PARALLEL COMPUTER ARCHITECTURES AND PARALLEL PROGRAMMING

To support one specific application with very high performance execution, the development of a special purpose hardware has always been a solution. As an example, in the field of VLSI design, hardware accelerators have been designed and marketed. Such hardware solutions deliver high performance, but they lack generality and flexibility. For this reason, in the following only universally programmable computer architectures will be considered.

To overcome some of the problems of the classical von Neumann execution model, alternative execution models have been proposed such as data flow machines, reduction machines, systolic arrays, neural networks etc. Since all of the alternative execution models are not compatible to the classical von Neumann machines, they found very little application. For this reason, they will not be considered, even though they partly support highly parallel computation.

Using the (modified) von Neumann computing model, two main forms of parallelism must be distinguished:

- Concurrency
- Pipelining.

Concurrent execution of the operations of an abstract machine means that more than one operation may be executed at the same time. No limitation on the respective beginning and end of the operations is implied with concurrency. Pipelined execution implies the subdivision of the operation into suboperations. Suboperations are executed strictly sequentially and fully synchronized by the stages of a pipeline. In this way, more than one operation of a pipelined machine may be executed, but with the individual suboperations in different stages of the pipeline.

Existing computer architectures may be viewed as hierarchies of abstract machines such as application-level machine, user-level machine, operating-system-level machine, machine-level machine, microprogram-level machine etc. The operation principles of every abstract machine may be organized sequentially or in parallel. When using parallelism, concurrency and pipelining may also be combined.

Regarding the application of parallelism, the following abstraction levels must be considered to characterize modern computer architectures:

- Program level
- Process (task, thread) level
- Groups-of-machine-instructions level
- Individual-machine-instruction level
- Parts-of-machine-instructions level.

Regarding the hardware of parallel computer architectures, there is no distinction between program- and process-level parallelism. The same applies for groups-of-machine-instructions level and individual-machine-instruction level. For that reason, a hardware-oriented architecture characterization distinguishes six classes of systems as indicated in table 1.

Table 1 Classes of parallelism

Parallelism level	Concurrency	Pipelining
Process Thread	**K** Multiprocessor Multicomputer (Net)	**K'** Macropipeline processor
Groups of instructions Individual instructions	**D** (Associative) Array processor	**D'** Instruction pipelining (Superscalar, VLIW)
Parts of instructions	**W** Parallel word processor	**W'** Machine instruction pipelining Arithmetical pipelining

Modern computer architectures mostly offer a combination of various forms of parallelism as listed in table 1. Händler, 1977, proposed a numerical description for computer architectures with the ECS (Erlangen Classification Scheme), grouping the degrees of parallelism for each of the 6 categories. In this way we obtain:

architecture classification = (K x K', D x D', W x W').

Examples for well known architectures would be:

Maspar MP/2 = (1 x 1, 16384 x 1, 4 x 1),
Intel iPSC/860 = (128 x 1, 1 x 2, 64 x 3).

The Maspar MP/2 is an array processor with a maximum of 16384 processing elements, each working with a wordlength of four bits. The Intel iPSC/860 is a multiprocessor with a maximum of 128 processors, each processor being an i860 microprocessor offering instruction pipelining (dual instruction mode), 64 bit wordlength and a maximum length of three for the arithmetical pipeline. The ECS description shows that the maximum parallelism for the array processor is in the number of processing elements (all being controlled by one instruction stream, disregarding space sharing). The maximum parallelism for the multiprocessor is in the number of processors or individual instruction streams being executed.

For the user, it is most important which forms of parallelism are exposed to the programmer and, therefore, must be explicitly programmed. Indeed, the virtualization of the hardware architecture is not realized for all forms of parallel machines as we are used from classical workstations. Instead, we have to distinguish between:

- Fully automatic parallelization

- Manual or interactive parallelization.

Fully automatic parallelization means that the parallelism of the hardware architecture is transparent for the application program. In that case, any program – even if developed for a sequential target architecture – is executable on the parallel target machine without modification of the source code. The generation of the parallel object code from the sequential source code is performed by some (optimizing) compiler and/or additional runtime hardware. In the case of manual parallelization, the programming language must allow for the explicit formulation of parallel constructs which may be either architecture-specific or independent. In the case of manual parallelization, most architectures offer tools to support the user in the task of parallelizing. We will call this case interactive parallelization.

Fully automatic parallelization is state-of-the-art for fine grain parallelism including the two lower levels of table 1 (parts-of-instruction level, individual-instruction level, and groups-of-instruction level, compare Karl, 1992). For the process and thread level, manual and at most interactive parallelization are available. Some exceptions apply to (virtual) shared memory machines, that will be described in the following. Parallelizing compilers for instruction pipelining, parallel word processing, machine instructions pipelining, and arithmetical pipelining are state-of-the-art. On the other hand, those techniques only offer limited parallelism due to details of the machine architecture and the algorithms. In the following, we will mainly concentrate on array processors, multiprocessors, multicomputers, and macropipeline processors because of their higher parallelism potential.

A final distinction has to be made between:

- Loosely coupled systems
- Tightly coupled systems.

In the literature, different definitions for those terms are given. We will use the term tightly coupled systems for architectures whose elements are grouped physically at very short distance: within a rack, or a set of racks. The term loosely coupled will be used, if the systems are interconnected to a local or wide area network, allowing for distances of at least several 100 meters between the different components. The interest in parallel systems started with the consideration of tightly coupled systems, because they offered a shorter communication time and more convenient way of programming. With the advent of faster interconnection schemes for computer networks such as FDDI or ATM, the development of optical media, and the availability of programming models for networks of homogeneous or heterogeneous computers such as PVM, MPI, PARMACS etc. (Sunderam, 1990), the differences between loosely and tightly coupled systems tend to disappear. This is mainly true for the consideration of applications where bandwidth is important (small number of big communication packages). For latency (big number of small communication packages), differences remain. Development of standards for message passing systems (MPI) will allow for portability of programs between loosely and tightly coupled parallel systems.

The classification of parallelism as shown in table 1 classifies parallel computer architectures by exclusively describing the computational section of the system. Main differences between architectures regarding the grouping of main memories, peripherals, and the interconnection structures are disregarded.

For loosely coupled systems, the interconnection network generally consists of one or a small number of a serial interconnection media shared by all members of the parallel system. Tightly coupled systems have various forms of interconnection networks. The main differences relate to:

- Parallelism of the individual interconnection line (serial, byte-, word-parallel)
- Interconnection structure (tree, mesh, pyramid, hypercube, ring etc.)
- Control and routing strategy (fixed versus switchable, single word versus packed, fixed versus adaptable routing, etc.).

A standard for building interconnection structures has not yet been found. For the user, the differences are mainly visible in terms of throughput, latency, and availability. In early systems (store-and-forward), even programming was affected. Modern systems tend to hide details of the interconnection network (virtually fully interconnected networks).

For the user, the arrangement of the elements of the main memory is more important, since it directly affects the programming model of the parallel system. Three main classes of systems may be found:

- Shared memory systems
- Distributed memory systems
- Virtual shared memory systems.

Systems with shared memory use one common main memory. Its address space may be accessed entirely by all processors. Therefore, communication between parallel program elements is possible by accessing a common memory element. The programming model is close to the one known from sequential systems. Existing programs can be easily ported to shared memory systems. A coarse grain load balancing algorithm may easily be implemented, if the scheduler of the operating system uses a central process waiting queue in memory accessible to all processors. Because of the fast shared memory communication, fine grain parallelism (distribution of the execution of individual instructions within a loop onto different processors) is possible. Automatic parallelization of this kind is often called symmetric multiprocessing (SMP) and realized by optimizing compilers. On the other hand, shared memory presents a global bottleneck for the system. Even if the work of the individual processors is decentralized through processor-specific caches, the number of processors must be restricted to at most 64 in order to avoid memory conflicts. We say that shared memory systems are not scalable (theoretically extendible to unlimited number of cooperating elements). If the performance available by such arrangements is sufficient for the application, shared memory systems are most convenient to use. Therefore, today most workstations, mainframes, and conventional supercomputers are available as shared memory multiprocessors (SEQUENT SYMMETRY, ALLIANT FX2800, SGI ONYX, CRAY YMP, etc.).

For distributed memory systems, the main memory is subdivided into memory modules attached to the individual processors. In such systems, the address space of the individual processor is restricted to its attached memory module. If access to shared data (physically located in a memory attached to a different processor) is needed, this access must be accomplished by an explicit communication call (send, receive). This explicit communication call is usually executed by some system hardware and software under supervision of the operating system. This type of explicit communication is called message passing communication. Various forms of message passing may be found (synchronous, asynchronous, broadcast, multicast, etc.). The message passing programming model implies that the user introduces message passing primitives in his program.

Therefore, porting existing software is more difficult as compared to shared memory systems. On the other hand, the bottleneck of a central main memory disappears, and such systems are theoretically fully scalable. In practice, the scalability is limited by the performance of the interconnection network regarding both bandwidth and latency. Automatic parallelization for message passing programming is only available for the SPMD-mode (Single Program Multiple Data) or data parallelism. Optimizing compilers such as High Performance FORTRAN offer this feature. For code parallelism, manual or interactive parallelization is necessary.

Distributed memory systems may either be tightly or loosely coupled. Recent work in the field of programming models for distributed memory systems tries to make disappear the differences between tightly and loosely coupled systems. Programming models known from loosely coupled systems (PVM) are being offered for tightly coupled systems, and models known from tightly coupled systems (NX) are being offered for loosely coupled systems. Of course, the use of loosely coupled homogeneous or heterogeneous computers is highly attractive for a number of reasons:

- The use of the machines is possible either as separate single systems or as a combined parallel system.
- The well known high percentage of idling workstations can be reduced.
- The large number of produced workstations reduces the price of the individual component for the parallel system.

For the realization of heterogeneous networks of computers, a number of problems must be solved such as the development of a single portable programming model, the reduction or the hiding of communication latency in the network, the development of scheduling and accounting techniques to supervise a large number of independent elements in a common environment, and the reliability and the dependability.

For distributed memory multiprocessors, a number of systems with up to several thousand processor nodes is commercially available. The main differences between the systems are visible in the physical realization of the interconnection structure. In most cases, the user programs virtual fully interconnected systems. Then, the messages to be sent are controlled by some communication processor building the interface to the interconnection network. The communication processor can either be realized as some separate hardware, or can be integrated into the main processor chip. The physical interconnection structure varies from machine to machine (mesh for PARAGON, fattree for CM-5, hierarchical clusters for Parsytec GC).

Programming for distributed memory multiprocessor still implies the use of architecture-specific libraries. For performance reasons, some systems offer both the functionality of parallel Unix (MACH or CHORUS) and low level message passing libraries (NX) for performance reasons.

Virtual shared memory systems try to combine the advantages of shared memory and distributed memory systems. Some software and/or hardware mechanism related to the memory management unit of the individual processors allows to use the shared memory programming model by simulating a shared memory on a physically distributed memory. If the mechanism offers enough efficiency, the advantages both in ease of programming and scalability are combined. The critical point is the implementation of the memory management unit and, once again, the performance of the interconnection network. Fully software implemented solutions as well as software/hardware implementations of virtual shared memory are available.

The first and only commercially available physical hardware implementation of such a machine is the KSR1. This machine disposes of a fully hardware implemented memory management unit deciding for each memory access whether the information is in local or non-local memory and fetching the information from remote memory if necessary. The processors of the KSR1 are clustered, up to 32 processors share a ring interconnection with a fixed access time. Several rings may be combined hierarchically by an additional interconnection ring. Access to a processor in a remote ring is slower than in the local ring. Following the principle of locality, the remote accesses should be less frequent. The suitability of virtual shared memory machines with large number of processors for different applications will be proved by users of the KSR and other machines that are announced.

The communication library LINDA, offering a virtual tuple space for shared data is a fully software implemented version of a virtual shared memory multiprocessor. LINDA is available for most of the well-known high-level programming languages and a number of tightly and loosely coupled systems both of the shared and distributed memory type.

In the remainder of this section, we will focus on programming coarse grain parallel systems. For the programmer, three aspects are important: programming languages, operating systems, and programming tools. Since some of the aspects of programming parallel systems are not independent of the hardware architectures, they have already been addressed in the previous paragraphs.

As for the hardware of parallel systems, standards for the software have not yet been established. Furthermore, the overall strategy of using parallel systems is discussed quite controversially and can be described by the following two extremes:

- For performance reasons, efficient programs are only realizable, if the programmer explicitly addresses all aspects of the parallel hardware in his program.
- For flexibility, ease of programming, and portability, all details of the parallel hardware should be fully transparent for the application.

As we will see in the following, there are good reasons for both standpoints. It is well-known, that explicitly designed parallel algorithms in general are more efficient than the parallelized versions of sequential algorithms. On the other hand, the lack of portability due to hardware-dependent programming is responsible for the small number of application codes for multiprocessor systems. Programs with dynamic process behavior (runtime-dependent) and the efficient use of systems for multiuser and multiprocess applications on the individual node require operating systems functionality with application-transparent scheduling etc. Therefore, in the future, systems hardware should be more transparent, but the user should design his algorithms with the explicit goal of exploiting the parallelism.

Regarding algorithms, three types of parallelization may be defined:

- Code partitioning
- Data partitioning
- Combination of code and data partitioning.

For numerical simulation, data partitioning and combinations of code and data-partitioning are most frequent (Michl, Maier, Wagner, Lenke, Bode, 1993).

Considering programming languages, all of the high level language paradigms (imperative such as FORTRAN, C, logical relational such as PROLOG, object-oriented directive such as SMALLTALK, functional applicative such as LISP) have been used for implicit or explicit parallelism. Implicit parallelism means that the user does not specify parallel constructs. Rather, some optimizing compiler produces code to address the parallelism of the hardware. Implicit parallelism is mainly available for fine grain parallel hardware. Explicit parallelism means, that the programmer specifies parallel constructs in his application program. Explicit parallelism mainly applies for coarse grain parallel hardware. Two types of offering explicit parallel language constructs may be distinguished:

- Extensions of existing programming languages by parallel libraries

- New parallel programming languages.

A typical case for the new programming language is OCCAM for the transputer. Such new languages, even if theoretically well designed and allowing for elegant programming styles are not yet well adopted because of their lack of compatibility. Therefore, especially in the field of numerical simulation, extensions of FORTRAN, C, and other well-known languages by message passing libraries are more frequent. Unfortunately, most of the message passing libraries are still architecture-dependent. In the future, architecture-independent communication libraries will allow for portability between different parallel systems. First approaches are PVM, MPI, PARMACS, and LINDA.

For operating systems, there is a main difference between shared memory and distributed memory implementations. Shared memory systems mostly offer some extended version of a standard operating system (mainly UNIX) with its full functionality. For distributed memory systems with a possibly large number of processing nodes, it first seemed not possible to implement the full operating system functionality on every node because of the necessity of replicating the full operating system code in every distributed memory module. Micro kernel techniques as offered in CHORUS and MACH and the availability of highly integrated memories have changed the situation since the beginning of the nineties. First distributed memory multiprocessors are available that offer the full functionality of an operating system including multiuser environment, virtual memory support etc. on the individual node (for example, MACH, OSF for PARAGON). In the future, distributed memory systems will probably offer a choice between a full operating system functionality for ease of programming and reduced node executive kernels for reasons of performance.

To support the development and testing process of parallel programs, a number of programming tools are needed: specification tools, mapping tools, debugger, performance analysis tools, visualizers etc. (compare Bemmerl, 1992). Shared memory systems mostly offer such tools as parts of their operating systems. Distributed memory systems first did not offer all of these functionalities. In the meantime, integrated hierarchical development tools for runtime or post-mortem analysis based on system monitoring and using a common graphical interface have been designed. Current research centers on application-oriented development tools and runtime-oriented high level tools for dynamic load balancing and fault tolerance (Bode, Dal Cin, 1993 and Ludwig, 1993).

SOME OPEN PROBLEMS

Unbalanced Computation to Communication Ratio

For shared memory systems, the timing of synchronization and communication to computation is well balanced, since both communication and synchronization are realized within a small number of instructions through the use of a shared memory access. For distributed memory systems with a message passing programming model, the execution times for computation and communication may differ by factors of 1000 and more. This is due to the fact that sent messages activate the operating system. The operating system and the underlying interconnection hardware and software have to packetize the information, establish a communication line between sender and receiver, transfer the information, and buffer the information, if the receiver is not available. For those systems, latency (time to start a message) is critical, bandwidth (quantity of information transfered per time unit, once the interconnection is established) is not. To circumvent this situation, a number of measures regarding algorithms, systems, and the implementation of the interconnection structure have been taken:

- The programmers should use algorithms with a minimum amount of communication/synchronization.
- If possible, the algorithms should be coded in a way that a large amount of information is packed into a small number of messages.
- The system software or interconnection hardware is packing messages sent to the same destination transparently to the application.
- Systems software hides the latency of communication by activating other processes.
- Development of interconnection systems that support a virtually fully interconnected system and perform the interconnection management on some hardware and/or software separate from the computing processor node (realized as a special purpose hardware on- or off-chip, or using an additional identical processor for communication).
- Enhancement of bandwidth by byte- or word-parallel transmission instead of bit-serial transmission between nodes.

All of the above-mentioned techniques have contributed to the reduction of latency and to the increase in bandwidth for tightly coupled coarse grain parallel systems. Loosely coupled systems such as networks of workstations have traditionally been using single bit serial interconnection media (ETHERNET). New transfer techniques such as FDDI, ATM increase bandwidths, but do not reduce latency. On the other hand, more parallel physical interconnection media (crossbar-switches) for the new interconnection media have been proposed.

To sum up: by using such techniques, parallel systems are getting higher communication power. Still, the user should search for algorithms that minimize communication and synchronization.

Portability in Programming

Portability is a major issue in computing. For the scientific community, it is sometimes deplorable that not the best technical solution dominates the hardware market, but the architecture for which the largest number of codes already exists. Regarding coarse grain parallel systems, once again a distinction has to be made between shared memory and distributed memory architectures.

For shared memory architectures, most parts of the parallel programs do not contain explicit parallel constructs. The only exceptions are language constructs for explicit parallelism, synchronization, and management. In general, the constructs are architecture-specific. Nevertheless, porting a program from one shared memory architecture to another is not very difficult.

For distributed memory architectures, a distinction has to be made between tightly coupled and loosely coupled systems. Since, for loosely coupled systems, homogeneous and heterogeneous types of systems are interconnected, the parallel programming models proposed have been implemented on a high software level (TCP sockets). Therefore, the different models proposed are not architecture-specific. The differences relate to the semantics of the communication constructs (synchronous, asynchronous, various forms of broadcast and collect) and the integration of development tools. A de-facto standard, used in a large variety of systems, even in tightly coupled systems, is PVM.

For tightly coupled distributed memory systems, as it was stated in the previous section, for performance reasons most architectures have highly machine-specific operating systems and programming models. Since, for communication calls, not only the identifier of the receiving process, but also the processor number of the receiver must be indicated, programs are even configuration-dependent. This means that a program developed for 16 processors will not run on a configuration of 32 processors of the same type of system without manual modification of the communication calls. Additionally, manual mapping of code and data is required (but in most cases not part of the communication library). The development of operating systems such as MACH and CHORUS with the full functionality of UNIX will support the way to portability. Additionally, programming models such as PVM and PARMACS, available on many machines, will allow for portability. The MPI (message passing interface), developed by the message passing interface forum, tries to impose a standard for message passing libraries.

For data partitioning or SPMD, the development of High Performance FORTRAN and the availability of compilers producing code for shared memory systems, for distributed memory multiprocessors, and for array processors as well as for sequential machines will help for portability of programs. Of course, the annotations made in High Performance FORTRAN to distribute code and data will always be targeted to some specific architecture and show less efficiency on other architectures.

All statements made about portability relate to source code portability. Object code portability seems not to be feasible for parallel systems in the near future.

PROGRAM DEVELOPMENT TOOLS FOR PARALLEL SYSTEMS

In the area of program development tools, differences exist in the availability of the components. Tools for the early development process, specification and verification as well as CASE tools are not available in commercial systems (Bemmerl, Bode, 1991). Tools for the late phases of the program development (debugger, performance analyzer, program flow, visualizer, mapper) are available, but their quality has to be enhanced regarding the support of application-specific demands and their scalability (support of massively parallel applications). Runtime-oriented tools offering application transparent dynamic load balancing (Ludwig, 1993), fault tolerance, support of various programming models (virtual shared memory) are in the development process but not yet available commercially. Such mechanisms can be integrated into parallel operating systems, enhancing their functionality. On the other hand, additional functionality of the operating system increases its size and decreases performance. Personally, I believe that the future will show systems with additional functionality.

CONCLUSIONS

Parallel systems are needed for performance and functionality. Flexibility, portability, availability, and dependability of such systems, especially scalable distributed memory architectures, are still small, but the way to better systems is open. For the user of parallel architectures, it is most important to remember that even the best parallel system will deliver poor efficiency, if the implemented algorithms do not exploit parallelism. Therefore, in the same way as future parallel systems have to be enhanced, parallel applications algorithms have to be developed. This is a main aim of the FORTWIHR consortium in the area of numerical simulation.

REFERENCES

[1] Bemmerl, T.: 1992, 'Programmierung skalierbarer Multiprozessoren', *BI Wissenschaftsverlag, Reihe Informatik* **Vol. 84**.

[2] Bemmerl, T., Bode, A.: 1991, 'An integrated environment for Programming Distributed Memory Multiprocessors', Bode, A. (ed.): 'Distributed Memory Computing', *LNCS* **Vol. 487**, pp. 130-142, 1991.

[3] Bode, A., Dal Cin, M.: 1993, 'Parallel Computer Architectures: Theory, Hardware, Software, Applications', Springer, LNCS, **Vol. 732**.

[4] Bromley, A.: 1992, 'Grand Challenges: High Performance Computing and Communication', *The FY 1992 US Research and Development Program.*

[5] Händler, W.: 1977, 'The impact of classification schemes on computer architecture', Proc. 1977 Conf. on Parallel Processing, pp. 7-15.

[6] Karl, W.: 1993, 'Parallele Prozessor-Architekturen', BI Wissenschaftsverlag, Reihe Informatik, **Vol. 93**.

[7] Ludwig, T.: 1993, 'Automatische Lastverwaltung für Parallelrechner', *BI Wissenschaftsverlag, Reihe Informatik*, **Vol. 94**.

[8] Sunderam, V.S.: 1990, 'PVM: A Framework for Parallel Distributed Computing', *Concurrency: Practice and Experience* **Vol. 2, No. 4**, pp. 315-339, Dec. 1990.

[9] Michl, T., Maier, S., Wagner, S., Lenke, M., Bode, A.: 1993, 'Dataparallel Navier-Stokes Solutions on Different Multiprocessors', In: Brebbia, Power (eds.), Applications of Supercomputers in Engineering III, pp. 263-277, Elsevier, 1993.

DESIGN OPTIMIZATION OF HIGH PERFORMANCE SATELLITES

R. Callies

Technische Universität München, Mathematisches Institut, FORTWIHR
D-80290 München, Germany

SUMMARY

Modern techniques of optimization and control considerably increase the performance of robotic satellites. As an example, a small Venus mission is presented for such a spacecraft. Not only a point-mass model is considered, but the full rigid body dynamics of a highly realistic model spacecraft is taken into account. The arising problems are formulated mathematically as boundary-value problems for complex systems of highly nonlinear differential equations. All scientific and technological constraints are exactly included as state and control constraints and interior point conditions. The numerical solution of the boundary-value problems is by a modified multiple shooting method. Problems of scaling and extremely small convergence areas require new solution techniques. For the first time the proof of mission feasibility is given. Design optimization leads to a cheap, robust satellite.

INTRODUCTION

Mathematics and Satellites – this is much more than the classical topic of trajectory optimization for a given space vehicle. Advanced methods of optimization in combination with powerful computers allow us now directly to interfere in the process of development and construction of a spacecraft. Trajectory optimization still is of fundamental importance, but it has to be seen in the context of the total system [1,5] and cannot be treated separately any longer.

What makes design optimization a challenge is that generally satellites are systems well optimized on the component level. A clear gain in overall performance is achieved only, if the optimization process strictly refers to the total system [6]. Although a single step improves system performace often only marginally, the sum of many tiny steps leads to improvements which make feasible new and demanding missions with technologies available today. And for every step the effects on and the interactions with the complete system – and its many subsystems – have to be considered. As an example: More powerful thrusters improve a satellite's performance by saving fuel. But every increase in thrust magnitude increases thruster mass, electric power consumption and the weight of the mechanical support structure. These effects very sensitively counteract the increase in system performance. Mass models, part selection and even the geometric placement of subsystems have to be included into the optimization process.

Mathematically this means that a new level of complexity of the models describing technical systems has to be handled in an efficient way. A small spacecraft to Venus is taken as an example to demonstrate important aspects of this optimal design approach.

With progress in microelectronics and -mechanics there is a renewed interest in those small, robotic spacecrafts for precisely defined and limited scientific missions: either stand-alone missions (e.g. to asteroids and comets) or support missions that significantly improve the scientific output of a main mission by additional measurements. These space systems are competitive especially for certain types of payloads: highly developed scientific instruments with a large fraction of state-of-the-art microelectronics (e.g. multi-colour high-resolution CCD-cameras), miniaturized components and time-critical experiments.

The advantage of such spacecrafts is the rapid and often cheaper access to space. Scientific flexibility is increased, development and engineering cycles are relatively short. But this is true only as long as strict limitations of the overall size and mass are observed and system complexity is kept low. On the other hand, every decrease in system performance will result in a further decrease of the payload fraction, which is rather low anyhow.

It is shown that modern techniques of optimization and control considerably improve the efficiency of that small Venus spacecraft. The optimization methods allow a simplification of the overall design by transfering complexity from the mechanical system – the "hardware" – to the build-in (electronic) intelligence and control. This leads to a cheap, robust satellite.

As for this, attention is mainly focused on the propulsion system which occupies by far the largest fraction of the system mass. The spacecraft is equipped with cheap thrusters. Maximum thrust level could be reduced by a factor of about 4. Thrusters are fixed to their position, no gimbals, no momentum wheels or other moving parts are needed – clearly an important factor for reliability. Momentum steering is used instead of the conventional thrust vector steering. To achieve this, thruster positions are optimized with the center of mass of the satellite off thrust axes. This operation mode has been fully optimized for the first time. New developed firing sequences of the thrusters reduce fuel consumption by more than 10 per cent.

Single stage design results in a further decrease in operation complexity and costs. Due to the special construction of the satellite, dual-mode operation is possible: long time spin-stabilized, short time three-axes-stabilized with a pointing accuracy of better 1°. By assistance of the despun antenna section even fine pointing is possible. All components on board are already found in commercial satellites.

To meet ESA restrictions the overall mass of the system is restricted to 250 kg, the outer diameter to 1280 mm and the total height of the satellite to 880 mm. This allows the spacecraft to be launched as a low cost piggy-back payload with the ARIANE launch vehicle. For this, the satellite is placed inside the adaptor between the two main payloads – typically two big commercial satellites. The performance of this space robot is demonstrated for an advanced mission profile.

THE MODEL SYSTEM

The numerical calculations are performed in the spherical ecliptic coordinate system $\mathbf{C} := (r, \varphi, \vartheta)$ with the Sun S at its origin. $\mathbf{X} := (X, Y, Z)$ denotes the corresponding cartesian coordinate system, $\mathbf{x} := (x_1, x_2, x_3)$ the body-fixed centroidal coordinate system and $\vec{\omega}$ the vector of the instantaneous angular velocity (Fig. 1).

THE MODEL SPACECRAFT The spacecraft is an improved version of the INEO-spacecraft [2]. For the mathematical treatment a careful abstraction is made from the very realistic engineering system without losing relevant information:
The spacecraft has the basic form of a decagonal prism. For mathematical treatment it is represented by a cylinder with an effective diameter of 1170 mm, an effective lenght of 860 mm and a uniform density. Initial mass is restricted to 250 kg. The amount of scientific instruments that can be carried has to be maximized.

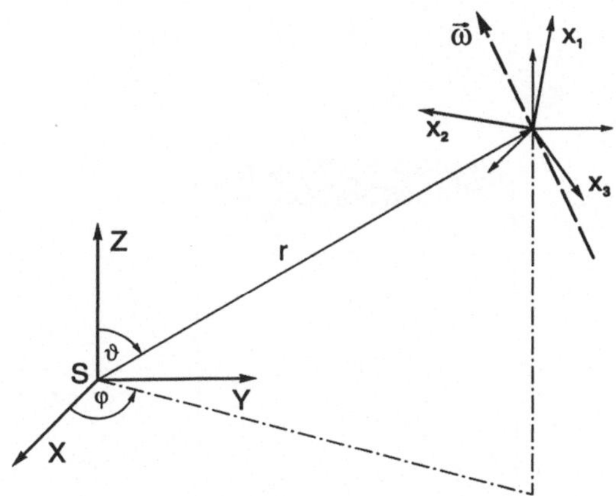

Fig. 1 Definition of the coordinate system: $\mathbf{C} := (r, \varphi, \vartheta)$ denotes the spherical inertial coordinate system, $\mathbf{x} := (x_1, x_2, x_3)$ the body-fixed centroidal coordinate system.

Two different types of thrusters are used: hydrazin-monopropellant thrusters and a solid rocket. All thrusters are fixed to the circular bottom of the spacecraft (see Fig. 2). The thrust direction of the following thrusters is parallel to the body-fixed rotation axis x_3: the solid rocket γ_0 (nominal thrust B_1 N, eff. thrust $B_1 \cdot f_1(B_1, B_2)$ N; eff. exhaust velocity $2880 \cdot f_1(B_1, B_2)$ m/s, centered, *one firing sequence!*), the two Hydrazin main engines γ_1 and γ_2 (nominal thrust B_3 N each, eff. thrust $B_3 \cdot f_2(B_3, B_4)$ N each; eff. exhaust velocity $2300 \cdot f_2(B_3, B_4)$ m/s, diametrical position with a mutual distance of 1160 mm) and the two Hydrazin control engines γ_3 and γ_4 (nominal thrust .5 N each, adjustable; eff. exhaust velocity 2300 m/s, diametrical position with a mutual distance of 1170 mm). For spin control two bidirectional Hydrazin thrusters γ_5 and γ_6 of the γ_3-type are used with their axes anti-/parallel to x_2.

B_1 to B_4 are free parameters subjected to optimization: B_1 and B_3 denote the thrust magnitudes in N, B_2 and B_4 the nozzle diameter of the respective thruster in mm. The specific impulse depends on the nozzle size and shape (functions f_1 and f_2). The following restrictions are obeyed: no thruster is installed on the satellite top (possible interference with experiments) and all attitude control engines are concentrated in two blocks. With reduced pointing accuracy the thrusters γ_3 and γ_4 can be omitted.

In the coordinate system defined above the movement of the spacecraft with its six degrees of freedom is given by the following highly non-linear system of differential equations of the general form $\dot{x} = f(x, u, t)$ (x vector of state variables, u vector of control variables; t independent variable, here: time). Equations of motion of the rigid body are taken from [3], equations of motion of the center of mass are derived in [4]:

$$\dot{r} = v_r \qquad \dot{\varphi} = \frac{v_\varphi}{r \sin \vartheta} \qquad \dot{\vartheta} = \frac{v_\vartheta}{r}$$

$$\dot{m} = -\frac{\gamma_0}{v_{A0}} - \frac{2|\gamma_5|}{v_{A5}} - \frac{\gamma_1 + \gamma_2}{v_{A1}} - \frac{\gamma_3 + \gamma_4}{v_{A3}}$$

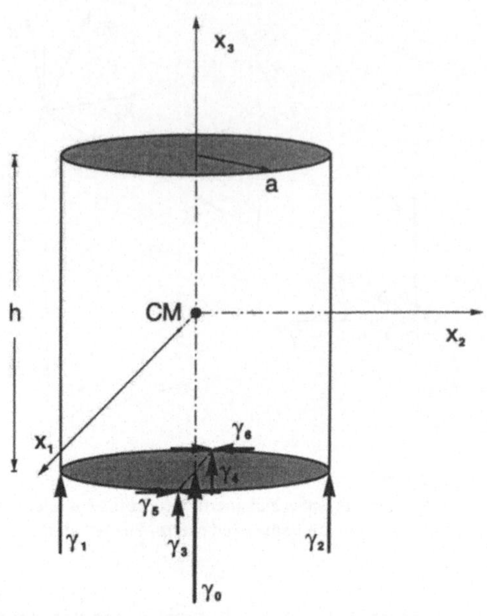

Fig. 2 Definition of the model spacecraft. CM denotes the center of mass, see text for further explanation.

$$\begin{aligned}
\dot{v}_r &= F_r + \frac{v_\varphi^2 + v_\vartheta^2}{r} - \frac{\tilde{\gamma} M}{r^2} + \frac{\xi}{mr^2} \\
&\quad -\tilde{\gamma} \sum_{j=1}^{n} \frac{m_j}{s_j^3} [r - r_j \cos\vartheta \cos\vartheta_j - r_j \sin\vartheta \sin\vartheta_j \cos(\varphi - \varphi_j)] \\
\dot{v}_\varphi &= F_\varphi - \frac{v_\varphi v_\vartheta}{r} - \frac{v_\varphi v_\vartheta}{r} \cdot \cot\vartheta - \tilde{\gamma} \sum_{j=1}^{n} \frac{m_j}{s_j^3} [r_j \sin\vartheta_j \sin(\varphi - \varphi_j)] \\
\dot{v}_\vartheta &= F_\vartheta - \frac{v_r v_\vartheta}{r} + \frac{v_\varphi^2}{r} \cdot \cot\vartheta \\
&\quad -\tilde{\gamma} \sum_{j=1}^{n} \frac{m_j}{s_j^3} [r_j \sin\vartheta \cos\vartheta_j - r_j \cos\vartheta \sin\vartheta_j \cos(\varphi - \varphi_j)]
\end{aligned} \tag{1}$$

$$\begin{aligned}
\dot{\omega}_1 &= -\frac{3a^2 - h^2}{3a^2 + h^2} \omega_2 \omega_3 + \frac{b \cdot (\gamma_2 - \gamma_1)}{m} \cdot \frac{12}{3a^2 + h^2} \\
\dot{\omega}_2 &= \frac{3a^2 - h^2}{3a^2 + h^2} \omega_1 \omega_3 + \frac{a \cdot (\gamma_4 - \gamma_3)}{m} \cdot \frac{12}{3a^2 + h^2} \\
\dot{\omega}_3 &= \frac{2a \cdot \gamma_5}{m} \cdot \frac{2}{a^2}
\end{aligned} \tag{2}$$

$$\begin{pmatrix} \dot{\beta}_0 \\ \dot{\beta}_1 \\ \dot{\beta}_2 \\ \dot{\beta}_3 \end{pmatrix} = \frac{1}{2} \begin{pmatrix} 0 & -\omega_1 & -\omega_2 & -\omega_3 \\ \omega_1 & 0 & \omega_3 & -\omega_2 \\ \omega_2 & -\omega_3 & 0 & \omega_1 \\ \omega_3 & \omega_2 & -\omega_1 & 0 \end{pmatrix} \cdot \begin{pmatrix} \beta_0 \\ \beta_1 \\ \beta_2 \\ \beta_3 \end{pmatrix}. \tag{3}$$

$\vec{r} := (r, \varphi, \vartheta)$ is the spacecraft's mass center location, $\vec{v} := (v_r, v_\varphi, v_\vartheta)$ its velocity and m the mass of the spacecraft. γ_i denotes the thrust magnitude of the respective engine and v_{Ai} its exhaust velocity.

$$\gamma_i := \begin{cases} B_1 \cdot f_1(B_1, B_2) \cdot \sin^2(\rho_i(t)) & i = 0 \\ B_3 \cdot f_2(B_3, B_4) \cdot \sin^2(\rho_i(t)) & i = 1, 2 \\ F_{i0} \cdot \sin^2(\rho_i(t)) & i = 3, 4 \\ F_{i0} \cdot \sin(\rho_i(t)) & i = 5, 6 \, . \end{cases} \tag{4}$$

For the exhaust velocity the following relationship holds:

$$v_{Ai} := \begin{cases} 2880 \cdot f_1(B_1, B_2) \; [m/s] & i = 0 \\ 2300 \cdot f_2(B_3, B_4) \; [m/s] & i = 1, 2 \\ 2300 \; [m/s] & i = 3, 4 \\ 2300 \; [m/s] & i = 5, 6 \, . \end{cases} \tag{5}$$

F_{i0} is the maximum nominal thrust and $\rho_i(t)$ the thrust control function ($i = 0 \ldots 6$). The following technical conditions have been considered in the state equations:

$$F_{03} = F_{04}, \; F_{05} = F_{06}, \; \rho_5(t) = \rho_6(t). \tag{6}$$

a and h are effective lengths of the spacecraft (see Fig. 2), $2b$ denotes the mutual distance of the thrusters γ_1 and γ_2.

$\vec{\omega}$ marks the instantaneous rotation axis of the spacecraft, the (redundant) Euler parameters $(\beta_0, \beta_1, \beta_2, \beta_3)$ describe the transformation between **x** and **X** in an unique way.

ξ denotes the solar pressure, $\tilde{\gamma}$ the gravity constant and M the mass of the Sun. The full complexity of the gravitational forces of many other celestial bodies ($n = 202$) like planets, moons or planetoids is included: m_j is the mass of the j-th body and $s_j(t) := |\vec{r}(t) - \vec{r}_j(t)|$ its distance from the spacecraft.

The resulting force of the thrusters

$$\vec{F} = (F_r, F_\varphi, F_\vartheta)^T := \vec{F}_0 + \vec{F}_r + \vec{F}_1 + \vec{F}_2 + \vec{F}_3 + \vec{F}_4, \tag{7}$$

measured in **C**, acts upon the center of mass of the spacecraft. The F_i are the solutions of the following equations

$$C(\beta_0, \beta_1, \beta_2, \beta_3) \cdot E(\varphi, \vartheta) \cdot \vec{F}_i = \vec{f}_i, \quad i \in \{r, 0, \ldots, 4\} \tag{8}$$

with

$$E = E(\varphi, \vartheta) := \begin{pmatrix} \cos(\varphi)\sin(\vartheta) & -\sin(\varphi) & \cos(\varphi)\cos(\vartheta) \\ \sin(\varphi)\sin(\vartheta) & \cos(\varphi) & \sin(\varphi)\cos(\vartheta) \\ \cos(\vartheta) & 0 & -\sin(\vartheta) \end{pmatrix} \tag{9}$$

describing the transformation between the reference frames **X** and **C** and

$$C = C(\beta_0, \beta_1, \beta_2, \beta_3) :=$$

$$\begin{pmatrix} \beta_0^2 + \beta_1^2 - \beta_2^2 - \beta_3^2 & 2(\beta_1\beta_2 + \beta_0\beta_3) & 2(\beta_1\beta_3 - \beta_0\beta_2) \\ 2(\beta_1\beta_2 - \beta_0\beta_3) & \beta_0^2 - \beta_1^2 + \beta_2^2 - \beta_3^2 & 2(\beta_2\beta_3 + \beta_0\beta_1) \\ 2(\beta_1\beta_3 + \beta_0\beta_2) & 2(\beta_2\beta_3 - \beta_0\beta_1) & \beta_0^2 - \beta_1^2 - \beta_2^2 + \beta_3^2 \end{pmatrix} \tag{10}$$

describing the transformation between the coordinate systems **X** and **x**.
The $\vec{f_i}$ are defined by

$$\begin{aligned}
\vec{f_0} &:= \frac{\gamma_0}{v_{A0}} \cdot (\omega_2 h/2, -\omega_1 h/2, v_{A0})^T \\
\vec{f_r} &:= \frac{\gamma_5}{v_{A5}} \cdot (\omega_2 h, -\omega_1 h, 0)^T \\
\vec{f_1} &:= \frac{\gamma_1}{v_{A1}} \cdot (\omega_2 h/2 - \omega_3 b, -\omega_1 h/2, \omega_1 b + v_{A3})^T \\
\vec{f_2} &:= \frac{\gamma_2}{v_{A1}} \cdot (\omega_2 h/2 + \omega_3 b, -\omega_1 h/2, -\omega_1 b + v_{A3})^T \\
\vec{f_3} &:= \frac{\gamma_3}{v_{A3}} \cdot (\omega_2 h/2, -\omega_3 a - \omega_1 h/2, \omega_2 a + v_{A3})^T \\
\vec{f_4} &:= \frac{\gamma_4}{v_{A3}} \cdot (\omega_2 h/2, \omega_3 a - \omega_1 h/2, -\omega_2 a + v_{A3})^T .
\end{aligned} \tag{11}$$

THE MASS MODEL For the Venus spacecraft the following mass model is used:

$$\begin{aligned}
m_{spacecraft} &= m_{payload} + m_{system} + m_{el.power}(1 + \delta_{el.power}) \\
&+ \sum_{i=0}^{6} m_{thruster,i}(1 + \delta_{thruster,i}) + \sum_{j=1}^{2} m_{fuel,j}(1 + \delta_{fuel,j})(1 + \Delta_{residual\, fuel,j}).
\end{aligned} \tag{12}$$

For the installation of an additional component with a mass of m_{comp} onboard the satellite (e.g. the installation of thruster i with a mass of $m_{thruster,i}$), additional support structure with a mass of $m_{comp,struc}$ is needed. $m_{comp,struc} = m_{comp,struc}(m_{comp})$ depends specifically on the component and nonlinearly on its mass m_{comp}; by definition [1]

$$\delta_{comp} := m_{comp,struc}/m_{comp}. \tag{13}$$

Experimental data for δ_{comp} vs. m_{comp} are published (see e.g. [5],[6]).

Electric power production is by solar cells (*Sol*). The mass of the solar power supply is described by

$$m_{el.power} = \alpha_1 \cdot P_{Sol}(t_0) + \alpha_2; \tag{14}$$

$$P_{Sol}(t_f) = P_{Sol}(t_0)\, e^{-\alpha_3 t_f} / r^{1.7} = g_1(m_{payload}) + g_2(\gamma_0, \dots \gamma_6) + \alpha_4 \tag{15}$$

is required as a side condition for the optimal solution. Index "0" denotes initial time, index "f" the *free* final time. $P_{Sol}(t)$ is the time-dependent electrical power output, the *local* constants α_j ($j = 1 \dots 4$) and the nonlinear functions g_l ($l = 1 \dots 2$) are derived from data published in literature (for details see below). α_3 is a fixed numerical value (engineering constant) describing system degradation.

A *local* representation of the mass of the i-th thruster is given by

$$m_{thruster,i} = \sum_{k=0}^{2} \alpha_{o,k,i} \gamma_{max,i}^{k} \,. \tag{16}$$

with maximum thrust $\gamma_{max,i}$. $\Delta_{residual\, fuel,j}$ denotes the part of fuel of type j that remains in the tank and cannot be used (about 1 – 2 %). Again the $\alpha_{0,k,i}$ and the $\Delta_{residual\, fuel,j}$ are determined from data collected from literature. The other subscripts should be self-explaining.

BOUNDARY CONDITIONS AND CONSTRAINTS The launch situation at time t_0 is prescribed ($\vec{r}(t_0)$ given, $\vec{v}(t_0)$ given, $m(t_0) = 1$). Final time t_f is free, here the following condition holds:

$$|\vec{r}(t_f) - \vec{r}_{Venus}(t_f)| \in [d_{V1}, d_{V2}], \quad d_{V1}, d_{V2} \in \mathbf{R} \text{ given.} \tag{17}$$

For scientific reasons, $d_{V2} - a_{Venus}$ is 250 km, whereas safety considerations yield $d_{V1} - a_{Venus}$ equal 100 km. a_{Venus} is the body radius of the Venus.

A minimum distance between the spacecraft and every celestial body is necessary because of safety reasons:

$$|\vec{r}(t) - \vec{r}_j(t)| \geq d_j (> 0), \ \ j = 1 \ldots n, \ \ \forall t \in [t_0, t_f]\,. \tag{18}$$

This leads to state constraints of 2nd order [7]. A similar condition restricts communication distance:

$$|\vec{r}(t) - \vec{r}_{Earth}(t)| \leq d_k \ \ \forall t \ \text{ and } \ t \in [t_0, t_f] \ \bigwedge \ (\gamma_1(t) > 0 \bigvee \gamma_2(t) > 0) \tag{19}$$

with d_k prescribed. The visibility condition can be written as

$$|\arccos[\vec{r}(t)\vec{r}_{Earth}(t)]/[|\vec{r}(t)||\vec{r}_{Earth}(t)|]| \geq \chi_0\,, \tag{20}$$

$\chi_0 = .223 \cdot \pi\,, \ \forall t$ with $t \in [t_0, t_f] \ \wedge \ (\gamma_1(t) > 0 \vee \gamma_2(t) > 0)$.

Control constraints are formulated in equation (4) - (6). Only one burn is allowed for the thruster γ_0.

MAXIMIZATION OF THE PAYLOAD

In a more abstract formulation the problem reads like this: Let us find a state function $x : \ [t_0, t_f] \longrightarrow \mathbf{R}^n$ and a control function $u : \ [t_0, t_f] \longrightarrow \mathrm{U} \subset \mathbf{R}^m$, which minimize the functional

$$\mathrm{I}(u) := -\, m_{payload} \tag{21}$$

subject to the conditions

$$\begin{aligned} \dot{x} &= f(x, u, t) = F(x, t) \cdot u \\ 0 &= g(t_0, x(t_0)) \in \mathbf{R}^n \\ 0 &= r(t_0, t_f, x(t_f)) \in \mathbf{R}^k, \ \ k < n \\ 0 &= q_i(t_i, x(t_i), x(t_i^-), x(t_i^+)) \in \mathbf{R}^l, \ \ l < n, \ i = 1, \ldots K \\ 0 &\leq C(x, t)\,. \end{aligned} \tag{22}$$

t_0 denotes the initial time, t_f the final time and t_i an intermediate time with an interior point condition. With thrusters fixed to their position, only linear controls (the time varying thrust magnitude of the main engines and the attitude control engines) occur. Only one firing sequence is allowed for the thruster γ_0; this restriction is integrated into the framework by an additional, piecewise constant control function.

The so defined problem of optimal control theory is transformed in a well-known manner (see e.g. [7],[8]) into a multi-point boundary value problem. There the following system of coupled nonlinear differential equations results ($H := \lambda^T f$):

$$\dot{x} = F(x, t) \cdot u \tag{23}$$

$$\dot{\lambda} = -H_x(x, \lambda, u, t). \tag{24}$$

In addition the Legendre-Clebsch condition has to be satisfied: $H_{uu}(x, \lambda, u, t)$ *pos. semidef.* The boundary conditions and interior point conditions are either prescribed a priori or obtained from the first variation of the extended functional. The solution of this boundary value problem satisfies the necessary conditions for an optimal solution.

The single components of a satellite are described by model functions and parameters that fit into this framework in a natural way. Only existing subsystems already space-qualified are taken into account. Scientific and technological constraints are included as state and control constraints and interior point conditions. The numerical solution of the boundary-value problems is by the multiple shooting method [9]-[11]. For the calculations a strongly modified version of the variant BOUNDSCO [11] is utilized. A high precision method is needed for the integration of the complicated systems of ordinary differential equations: extrapolation methods have been chosen for this purpose to solve the arising initial value problems [12]. For these complex problems arising from astronautics additional numerical techniques [4],[13] have been developed and applied.

Scaling problems and extremely small convergence areas required new solution techniques:

– *Interior points* T_i , $i = 1, \ldots, 14$, $T_i < T_j$ for $i < j$, $T_i \in [t_0, t_f]$ $\forall i$ are additionally introduced and allow the rescaling of the independent variable t in sections: $\zeta_i := e_{i,1}t + e_{i,0}$ $\forall t \in]T_{i-1}, T_i]$, $i = 1, \ldots, 15$ and $T_0 := t_0$, $T_{15} := t_f$. The scaling factors $e_{i,k}$ are determined in every iteration step of the multiple shooting method in that way, that $cond(M) \leq MIN * 1.5$; here M denotes the multiple shooting matrix [9] and MIN is an approximation of the minimum condition number that can be obtained by scaling operations.

– For the determination of the system functions (m_*, δ_*, g_*) C^∞-functions are semi-automatically fitted to experimental date published in literature (see e.g. [5]). In order to preserve numerical stability and to make calculations more effective, most of these functions are then interpolated by linear or cubic polynoms in the neighborhood of the current approximation of the solution (*local* interpolation). The α_i, $\alpha_{0,k,i}$ in (12-16) are achieved in this way. This interpolation is – unlike a static interpolation – actualized and improved *in every iteration step* of the multiple shooting method. Reduction of relative accuracy in the functional $I(u)$ is less than 10^{-9}.

– It is not before the last few iteration steps that the full problem is precisely calculated. Before that, the spiraling-up maneuver and the interplanetary trajectory are handled separately (scaling problems arise due to different time and lenght scales), the rigid body motion and the motion of the center of mass are calculated sequentially: In the early steps of calculation the respective other parts are handled only approximately.

THE EXAMPLE MISSION

As an example, a flyby mission at the Venus has been designed as a reference mission. Fig. 3 (Earth escape maneuver) and Fig. 4 (interplanetary flight) show a fully optimized flight trajectory from the Earth to the Venus.

With an Ariane launch the spacecraft is delivered into an GTO-orbit with a perigee of 185 km and an apogee of 35786 km. Until now no other precise informations concerning launch are available. Therefore, a standard situation is defined: At launch time $T0$ the spacecraft has a distance of 125000 km from Earth, and the hyperbolic launch velocity $\vec{V}_{HL}$ lies in the plane of ecliptic and is parallel to the velocity vector of the Earth at $T0$ (*Definition* of launch geometry and $T0$!). L is the position of the spacecraft at $t = T0$. The 125000 km-boundary is artificially introduced, but has *no* mathematical meaning except the change of the coordinate system for technical reasons. The two coordinate systems are coupled by an interior point condition. V_{HL} is optimized; optimization of its direction will further decrease fuel consumption [13]. The total final trajectory (spiraling-up and interplanetary flight) was calculated without splitting-up or simplification.

Fig. 3 shows the projection of the optimal flight trajectory of the spacecraft to the base

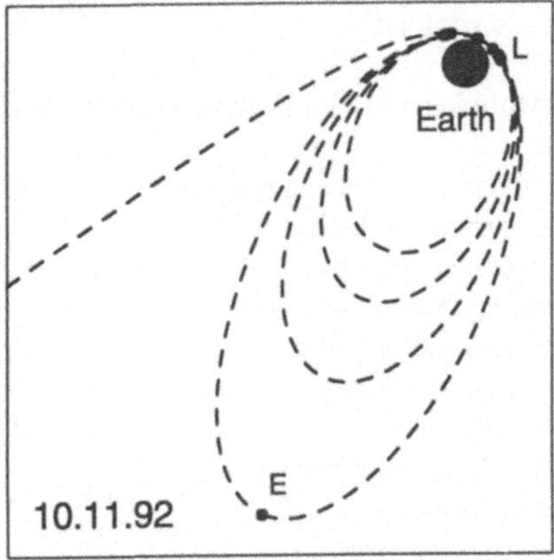

Fig. 3 Projection of the three-dimensional Earth escape maneuver. Mass consumption 87.9 kg, time for spiraling out 4 days.

plane of a geocentric coordinate system. Solid lines mark those parts of the trajectory where thrusters are on, broken lines those with the thrust off. First thrust arc starts near the perigee of the GTO-ellipse on November 6th, 1992. After 4 orbits and slightly more than 4 days the spacecraft has reached its optimal escape velocity. Launch time $T0$ is November 10th, 1992; the spiraling-up from the GTO-ellipse takes 87.9 kg of fuel. Optimal hyperbolic escape velocity is 2.71 km/s. On the coast arcs the spacecraft is spin-stabilized with 5 rpm and the spacecraft main axis x_3 has to be perpendicular to the plane of motion. Three axes stabilization is necessary only while thrusters are in operation. The special design of the spacecraft allows this fuel-saving dual-mode operation (spin- and three-axes-stabilization).

Two important features of the optimal spiraling-up maneuver should be mentioned. The solid rocket is fired not before the last turn. This allows effectively to use the gravity effect. At the same time it becomes possibly to use a fixed and relatively small solid booster for a large variety of launching situations with only very little loss in overall performance. The single burn of the solid booster with its higher specific impulse is fully utilized, the velocity increment additional necessary and varying with the special situation is produced by the Hydrazin thrusters. The solid rocket has a usable mass of fuel of 40 kg. An improved efficiency of the solid rocket is achieved by an additional thrust arc E near the apogee of the last turn around the Earth. Here the spacecraft is repositioned optimal for the escape burn.

Fig. 4 shows the three-dimensional plot of the complete and optimal interplanetary flight trajectory. The orbits of Earth and Venus are indicated by solid lines. The only deep space maneuver takes place on April 12th, 1993 (fuel consumption 71.5 kg), the thrusters γ_1 and γ_2 are firing for about 2 hours. The spacecraft reaches Venus on December 22th, 1993. The positions of Earth at the departure date D (open circle) and the positions of Venus and Earth

Table 1: The spacecraft to Venus

Initial mass m_0	:	250.0 kg
Fuel	:	162.0 kg
– Solid	:	40.0 kg
– Hydrazin Mono	:	122.0 kg
Scientific payload	:	24.0 kg
Altitude and Orbit Control System (AOCS)		
Solid booster γ_0 (1 firing sequence)	:	300 N nom.
– nom. exhaust velocity	:	2880 m/s
– eff. exhaust velocity	:	2868 m/s
2 Mono-Hydrazin thrusters γ_1, γ_2	:	10 N nom.
– nom. exhaust velocity	:	2300 m/s
– eff. exhaust velocity	:	2296 m/s
– mutual distance	:	1160 mm
4 Mono-Hydrazin thrusters $\gamma_3 \ldots \gamma_6$	:	0.5 N
– exhaust velocity	:	2300 m/s
The Optimal Trajectory		
Launch date	:	6.11.1992
Earth escape maneuver	:	10.11.1992
Deep space maneuver	:	12.4.1993
Arrival date	:	22.12.1993
Fuel consumption		
– total	:	159.4 kg
– subtotal solid	:	40.0 kg
– subtotal Hydrazin	:	119.4 kg
– for Earth escape	:	87.9 kg
– for deep space maneuver	:	71.5 kg
Earth escape velocity	:	2.71 km/s
Venus flyby velocity	:	3.29 km/s
Important Aspects		
effective momentum steering with fixed thrusters		
new operation mode for the solid rocket		
low thrust level, cheap design		
compact, fault-tolerant design		
dual-mode operation: spin- and three-axes-stabilized		

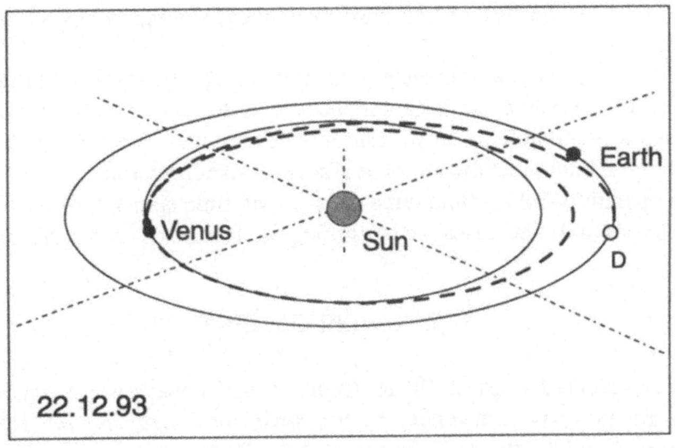

Fig. 4 Three-dimensional plot of the swingby at Venus. Departure from Earth orbit on November 10th, 1992, arrival at Venus on December 22th, 1993; $m(t_f) = .362$.

at the arrival date at Venus (filled circles) are plotted in Fig. 4. Total fuel consumption for the journey to Venus is 159.4 kg. On the coast arcs of the interplanetary flight the spacecraft is spin-stabilized with 5 rpm and the main axis x_3 has to be perpendicular to the plane of ecliptic. Table 1 summarizes the relevant data of the optimized spacecraft. The unfavourable launch date is compensated by a longer total flight time and one and a half turns around the Sun. The deep space maneuver allows to widen the trajectory so that Venus can overtake the spacecraft; at the first orbit the space probe has reached the meeting point long before Venus. The trajectory lies very close to the plane of ecliptic, the spacecraft meets Venus near the intersection between the plane of ecliptic and the plane of motion of the Venus.

SUMMARY AND CONCLUSIONS

Modern techniques of optimization and control considerably increase the performance of robotic satellites. As an example, a small Venus mission is presented for such a spacecraft. The full rigid body dynamics of a highly realistic model spacecraft is taken into account. The arising problems are formulated mathematically as boundary-value problems for complexe systems of highly nonlinear differential equations. All scientific and technological constraints are exactly included as state and control constraints and interior point conditions. The numerical solution of the boundary-value problems is by a modified multiple shooting method. Problems of scaling and extremely small convergence areas are overcome by new solution techniques.

The special reformulation of the problem with redundant variables and varying accuracies in the approximation of model parameters makes it well suited for the use with modern computers, especially parallel processing.

For the first time the proof of mission feasibility is given. Design optimization leads to a cheap, robust satellite. Full redundance against one failure is given, in the thruster area a twofold redundance is achieved. Optimal momentum steering is shown to be a competitive operation mode. The small losses in fuel in comparison with thrust vector steering are overcompensated by the decrease in weight because of a simpler mechanical system. Additionally an increase of reliability is achieved.

The special developed firing sequence of the thrusters (solid thruster burn on the last orbit around Earth after an apogee repositioning maneuver) reduces fuel consumption by more than 10 per cent. Thus it is possible to do without more expensive bipropellant thrusters or a two stage design that would not meet the required mass and size limitations.

Dual mode operation – short time three-axes- , long time spin-stabilized – is proven to be possible. This considerably increases the scientific possibilities of such a mission.

ACKNOWLEDGEMENT

The author is indebted to Prof. Dr. R. Bulirsch who always encouraged and supported this work. This research has been funded by the *Bavarian Consortium on High Performance Scientific Computing* and by the *DFG* in the *Special Research Center on Transatmospheric Vehicles (SFB 255)*.

REFERENCES

[1] Callies, R., *Optimal design of a mission to Neptune.* In: Optimal Control – Calculus of Variations, Optimal Control Theory and Numerical Methods. Eds.: R. Bulirsch, A. Miele, J. Stoer, K.H. Well. International Series of Numerical Mathematics, Vol. 111, Birkhäuser Verlag, Basel (1993), S. 341–349.

[2] Iglseder, H., Arens-Fischer, W., Keller, H.U., Arnold, G., Callies, R., Fick, M., Glassmeier, K.H., Hirsch, H., Hoffmann, M., Rath, H.J., Kührt, E., Lorenz, E., Thomas, N., Wäsch, R., *INEO – Imaging of Near Earth Objects.* COSPAR-Paper 1-M.1.03, Washington (1992).

[3] Junkins, J.L., Turner, J.D., *Optimal Spacecraft Rotational Maneuvers*, Elsevier Science Publ. Comp., New York, 1986.

[4] Bulirsch, R., Callies, R., *Optimal trajectories for a multiple rendezvous mission to asteroids.* Acta Astronautica **26** (1992) 587-597.

[5] Larson, W.J., Wertz, J.R. (ed.), *Space Mission Analysis and Design*, Kluwer Academic Publisher, Dordrecht 1992.

[6] Ruppe, H.O., *Introduction to Astronautics*, Academic Press, New York 1966.

[7] Bryson, A.E., Ho, Y.-C., *Applied Optimal Control*, Revised Printing, Hemisphere Publishing Corp., Washington D.C., 1975.

[8] Oberle, H.J., *Numerical Computation of Minimum-Fuel Space-Travel Problems by Multiple Shooting*, Report TUM-MATH-7635, Dept. of Mathematics, Munich Univ. of Technology, Germany, 1976.

[9] Bulirsch, R., *Die Mehrzielmethode zur numerischen Lösung von nichtlinearen Randwertproblemen und Aufgaben der optimalen Steuerung*, Report of the Carl-Cranz-Gesellschaft e.V., Oberpfaffenhofen, 1971.

[10] Deuflhard, P., *Ein Newton-Verfahren bei fast singulärer Funktionalmatrix zur Lösung von nichtlinearen Randwertaufgaben mit der Mehrzielmethode*, Thesis, Cologne, 1972.

[11] Oberle, H.J., *Numerische Behandlung singulärer Steuerungen mit der Mehrzielmethode am Beispiel der Klimatisierung von Sonnenhäusern*, Thesis, Munich, 1977.

[12] Bulirsch, R., Stoer, J., *Numerical Treatment of Ordinary Differential Equations by Extrapolation Methods*, Num. Math. **8** (1966) 1-13.

[13] Callies, R., *Optimale Flugbahnen einer Raumsonde mit Ionentriebwerken*, Thesis, Munich, 1990.

NUMERICAL SIMULATIONS OF DYNAMICAL GINZBURG-LANDAU VORTICES IN SUPERCONDUCTIVITY

Z. Chen[1], K.-H. Hoffmann
Lehrstuhl für Angewandte Mathematik
Technische Universität München
Dachauerstr. 9a, 80335 München, Germany.

SUMMARY

In this paper, by using a semi-implicit finite element scheme proposed and analyzed previously by the authors, we simulate numerically the motion of vortices under the influence of the applied magnetic field in superconductivity via solving the time-dependent Ginzburg-Landau equations with the Lorentz gauge $\mathrm{div}\mathbf{A} + \phi = 0$.

INTRODUCTION

The phenomenological Ginzburg-Landau complex superconductivity model is designed to describe the phenomenon of vortex structure in the superconducting-normal phase transitions. The time-dependent Ginzburg-Landau (TDGL) model derived by Gor'kov and Éliashberg [9] from averaging the microscopic BCS theory offers a useful starting point in studying the dynamics of superconductivity. Let Ω be a bounded domain in $\mathbb{R}^2$ and $(0,T)$ the time interval. Denote by $Q = \Omega \times (0,T)$ and $\Gamma = \partial\Omega \times (0,T)$. The TDGL model can be formulated as in the following system of nonlinear partial differential equations

$$\frac{\hbar^2}{2m_s} D^{-1}\Big(\frac{\partial}{\partial t} + i\frac{e_s}{\hbar}\phi\Big)\psi + \big(\alpha + \beta|\psi|^2\big)\psi + \frac{1}{2m_s}\Big(\frac{\hbar}{i}\nabla - \frac{e_s}{c}\mathbf{A}\Big)^2\psi = 0, \quad \text{in } Q \qquad (1)$$

$$\frac{c}{4\pi}\mathbf{curl}\mathrm{curl}\mathbf{A} = -\sigma\Big(\frac{1}{c}\frac{\partial \mathbf{A}}{\partial t} + \nabla\phi\Big) + \frac{e_s}{m_s}\Re\Big[\Big(\frac{\hbar}{i}\nabla\psi - \frac{e_s}{c}\mathbf{A}\psi\Big)\bar{\psi}\Big], \quad \text{in } Q \qquad (2)$$

with the boundary and initial conditions

$$\Big(\frac{\hbar}{i}\nabla\psi - \frac{e_s}{c}\mathbf{A}\psi\Big)\cdot\mathbf{n} = 0, \ \mathrm{curl}\mathbf{A} = H, \quad \text{on } \Gamma \qquad (3)$$

$$\psi(x,0) = \psi_0(x), \ \mathbf{A}(x,0) = \mathbf{A}_0(x), \quad \text{on } \Omega \qquad (4)$$

where $\mathbf{n} = (n_1, n_2)$ denotes the unit outer normal of the boundary $\partial\Omega$, $\Re[\,\cdot\,]$ denotes the real part of the quantity in the brackets $[\,\cdot\,]$, and curl, **curl** denote the curl operators on $\mathbb{R}^2$ defined by

$$\mathrm{curl}\mathbf{B} = \frac{\partial B_2}{\partial x_1} - \frac{\partial B_1}{\partial x_2}, \qquad \mathbf{curl}\zeta = \Big(\frac{\partial\zeta}{\partial x_2}, -\frac{\partial\zeta}{\partial x_1}\Big)^T.$$

[1]Present address: Institute of Mathematics, Academia Sinica, Beijing 100080, People's Republic of China.

This work was supported by the Deutsche Forschungsgemeinschaft, SPP "Anwendungsbezogene Optimierung und Steuerung".

Here ψ is a complex valued function and is usually referred to as the order parameter so that $|\psi|^2$ gives the relative density of the superconducting electron pairs; $\bar{\psi}$ is the complex conjugate of ψ; $\mathbf{A}$ is a real vector potential for the total magnetic field; ϕ is a real scalar function called the electric potential; H is the applied magnetic field, viewed as a vector, the magnetic field points out of the (x_1, x_2)-plane; $\alpha < 0$ and $\beta > 0$ are constants; c is the speed of light; e_s and m_s are the charge and mass, respectively, of the superconducting carriers; $2\pi\hbar$ is Planck's constant; and D and σ are the normal state diffusion constant and conductivity, respectively.

There have been several numerical attempts to solve the TDGL equations in the physical literature (cf. e.g. [6], [7], [8], [10]). In these computations, the dependence of the system on the electric potential ϕ is eliminated via a gauge transformation such that $\phi = 0$. In that context, the system looks simpler but the equation involving the magnetic potential $\mathbf{A}$ is no longer coercive in $\mathbf{H}^1(\Omega)$, which in turn results in some difficulties in designing numerically convergent schemes for the TDGL equations (see the discussion and analysis in [5] for this gauge choice). In [2], by choosing the gauge transformation such that

$$\phi = -\frac{c}{4\pi\sigma}\mathrm{div}\mathbf{A} \ \text{ in } Q \quad \text{and} \quad \mathbf{A}\cdot\mathbf{n} = 0 \ \text{ on } \Gamma \tag{5}$$

the authors proposed and analyzed a semi-implicit finite element scheme for solving the TDGL model (1)-(4). This method can be easily implemented in the practical computations. The purpose of this paper is to report the numerical results obtained by using this method. We also remark that the existence and uniqueness of strong solutions to the system (1)-(4) under the gauge choice (5) has been proved in [3].

The layout of the paper is following: in next section we describe briefly the approximate method in [2]; then, we report several numerical experiments to show the effectiveness of the method in simulating the motion of vortices in superconductors.

In the remainder of this section we rescale the system of equations (1)-(4) with the gauge relation (5) in order to simplify the notations. Let the magnetic penetration depth λ and the coherence length ξ be defined by

$$\lambda = \sqrt{-\frac{\beta m_s c^2}{4\pi\alpha e_s^2}}, \quad \xi = \sqrt{-\frac{\hbar^2}{2m_s\alpha}},$$

and the Ginzburg-Landau parameter $\kappa = \lambda/\xi$. The equations (1)-(4) with the gauge relation (5) can be rescaled to measure x in units of λ, t in units of t_0, $\mathbf{A}$ in units of $\sqrt{2}\lambda H_c$, ψ in units of ψ_c, ϕ in units of $\hbar\kappa/e_s t_0$, H in units of $\sqrt{2}\lambda H_c$, where (see [6]) $\psi_c = \sqrt{-\alpha/\beta}, H_c = \sqrt{4\pi\alpha^2/\beta}$ and $t_0 = 4\pi\lambda\sigma^2/c^2 = \xi^2/12D$, to obtain, in normalized dimensionless units,

$$\eta\frac{\partial\psi}{\partial t} - i\eta\kappa\mathrm{div}\mathbf{A}\psi + (|\psi|^2 - 1)\psi + \left(\frac{i}{\kappa}\nabla + \mathbf{A}\right)^2\psi = 0, \quad \text{in } Q \tag{6}$$

$$\frac{\partial\mathbf{A}}{\partial t} - \Delta\mathbf{A} + \Re\left[\left(\frac{i}{\kappa}\nabla\psi + \mathbf{A}\psi\right)\bar{\psi}\right] = 0, \quad \text{in } Q \tag{7}$$

$$\left(\frac{i}{\kappa}\nabla\psi + \mathbf{A}\psi\right)\cdot\mathbf{n} = 0, \ \mathrm{curl}\mathbf{A} = H, \quad \text{on } \Gamma \tag{8}$$

$$\psi(x,0) = \psi_0(x), \ \mathbf{A}(x,0) = \mathbf{A}_0(x), \quad \text{on } \Omega. \tag{9}$$

Here, we have substituted the following dimensionless form of the gauge relation (5)

$$\phi = -\mathrm{div}\mathbf{A} \ \text{ in } Q \quad \text{and} \quad \mathbf{A}\cdot\mathbf{n} = 0 \ \text{ on } \Gamma$$

into the normalized form of (1)-(4). The rescaled system (6)-(9) is the starting point of the numerical simulations reported in this paper.

THE DISCRETE PROBLEM

We first introduce the weak formulation of the problem (6)-(9). For any integer $m \geq 1$, let $\mathcal{H}^m(\Omega)$ denote the Sobolev space of complex valued functions having their real and imaginary part in $H^m(\Omega)$ and $\mathbf{H}^m(\Omega)$ denote the Sobolev space of vector valued functions having their components in $H^m(\Omega)$. Let $\mathbf{H}^1_n(\Omega) = \{\mathbf{A} \in \mathbf{H}^1(\Omega) : \mathbf{A} \cdot \mathbf{n} = 0 \text{ on } \partial\Omega\}$ and

$$\mathcal{W} = L^2(0,T;\mathcal{H}^1(\Omega)) \cap H^1(0,T;(\mathcal{H}^1(\Omega))'),$$
$$\mathbf{W}_n = L^2(0,T;\mathbf{H}^1_n(\Omega)) \cap H^1(0,T;(\mathbf{H}^1_n(\Omega))'),$$

where $(\mathcal{H}^1(\Omega))'$ and $(\mathbf{H}^1_n(\Omega))'$ are the dual spaces of the Banach spaces $\mathcal{H}^1(\Omega)$ and $\mathbf{H}^1_n(\Omega)$, respectively. The weak formulation for the system (6)-(9) is then to find $(\psi, \mathbf{A}) \in \mathcal{W} \times \mathbf{W}_n$ such that

$$\psi(x,0) = \psi_0(x), \; \mathbf{A}(x,0) = \mathbf{A}_0(x) \tag{10}$$

and

$$\begin{aligned}
&\eta \int_0^T \int_\Omega \frac{\partial \psi}{\partial t} \omega \text{ dxdt} - i\eta\kappa \int_0^T \int_\Omega \text{div}\mathbf{A}\psi\omega \text{ dxdt} \\
+ &\int_0^T \int_\Omega \Big(\frac{i}{\kappa}\nabla\psi + \mathbf{A}\psi\Big)\Big(-\frac{i}{\kappa}\nabla\omega + \mathbf{A}\omega\Big) \text{ dxdt} \\
+ &\int_0^T \int_\Omega (|\psi|^2 - 1)\psi\omega \text{ dxdt} = 0, \quad \text{for any } \omega \in L^2(0,T;\mathcal{H}^1(\Omega))
\end{aligned} \tag{11}$$

$$\begin{aligned}
&\int_0^T \int_\Omega \frac{\partial \mathbf{A}}{\partial t}\mathbf{B} \text{ dxdt} + \int_0^T \int_\Omega \Big[\text{curl}\mathbf{A}\text{curl}\mathbf{B} + \text{div}\mathbf{A}\text{div}\mathbf{B}\Big] \text{ dxdt} \\
+ &\int_0^T \int_\Omega \Re\Big[\Big(\frac{i}{\kappa}\nabla\psi + \mathbf{A}\psi\Big)\bar{\psi}\Big]\mathbf{B} \text{ dxdt} \\
= &\int_0^T \int_{\partial\Omega} H(\mathbf{B}\cdot\boldsymbol{\tau}) \text{ ds dt}, \quad \text{for any } \mathbf{B} \in L^2(0,T;\mathbf{H}^1_n(\Omega))
\end{aligned} \tag{12}$$

where $\boldsymbol{\tau} = (-n_2, n_1)$ is the unit tangent to $\partial\Omega$.

We shall make use of backward Euler scheme to discretize the problem (10)-(12) in time. Let M be a positive integer and $\Delta t = T/M$ be the time step. For any $n = 1, 2, \cdots, M$, we define $t^n = n\Delta t$ and $I^n = (t^{n-1}, t^n]$. Furthermore, we denote $\partial\eta^n = (\eta^n - \eta^{n-1})/\Delta t$ for any given sequence $\{\eta^n\}_{n=0}^M$ and $\eta^n = \eta(\cdot, t^n)$ for any given function $\eta \in C(0,T;X)$ with some Banach space X.

In space we shall utilize finite element approximations. Let $\{\Delta_h\}_{h>0}$ be a family of quasi-uniform triangulations of Ω. For convenience we shall assume Ω is a convex polygonal domain in $\mathbb{R}^2$ and thus we can assume that $\Omega = \Omega_h = \cup_{K\in\Delta_h} K$. Let $\mathcal{S}_h \subset \mathcal{H}^1(\Omega)$ and $\mathbf{V}_h \subset \mathbf{H}^1_n(\Omega)$ be the finite element spaces which satisfy the following approximation properties

$$\|\zeta - \zeta_\text{I}\|_{H^j(\Omega)} \leq Ch^{2-j}\|\zeta\|_{H^2(\Omega)}, \; \forall\zeta \in \mathcal{H}^2(\Omega), \; j = 0, 1$$
$$\|\mathbf{Q} - \mathbf{Q}_\text{I}\|_{H^j(\Omega)} \leq Ch^{2-j}\|\mathbf{Q}\|_{H^2(\Omega)}, \; \forall\zeta \in \mathbf{H}^2(\Omega), \; j = 0, 1$$

where $\zeta_I \in \mathcal{S}_h$ and $\mathbf{Q}_I \in \mathbf{V}_h$ denote the interpolate of $\zeta \in \mathcal{C}(\bar{\Omega})$ in $\mathcal{S}_h$ and $\mathbf{Q} \in \mathbf{C}(\bar{\Omega})$ in $\mathbf{V}_h$, respectively. The discrete problem in [2] is then, for $n = 1, 2, \cdots, M$, to find $(\psi_h^n, \mathbf{A}_h^n) \in \mathcal{S}_h \times \mathbf{V}_h$ such that

$$\psi_h^0 = (\psi_0)_I, \quad \mathbf{A}_h^0 = (\mathbf{A}_0)_I, \tag{13}$$

and

$$\begin{aligned} & \eta \int_\Omega \partial\psi_h^n \omega \,\mathrm{dx} - i\eta\kappa \int_\Omega \mathrm{div}\mathbf{A}_h^n \psi_h^n \omega \,\mathrm{dx} \\ + & \int_\Omega \Big(\frac{i}{\kappa}\nabla\psi_h^n + \mathbf{A}_h^n\psi_h^n\Big)\Big(-\frac{i}{\kappa}\nabla\omega + \mathbf{A}_h^n\omega\Big)\,\mathrm{dx} \\ + & \int_\Omega (|\psi_h^n|^2 - 1)\psi_h^n \omega \,\mathrm{dx} = 0, \quad \text{for any } \omega \in \mathcal{S}_h \end{aligned} \tag{14}$$

$$\begin{aligned} & \int_\Omega \partial\mathbf{A}_h^n \mathbf{B} \,\mathrm{dx} + \int_\Omega \Big[\mathrm{curl}\mathbf{A}_h^n \mathrm{curl}\mathbf{B} + \mathrm{div}\mathbf{A}_h^n \mathrm{div}\mathbf{B}\Big]\,\mathrm{dx} \\ + & \int_\Omega \Re\Big[\Big(\frac{i}{\kappa}\nabla\psi_h^{n-1} + \mathbf{A}_h^n\psi_h^{n-1}\Big)\bar{\psi}_h^{n-1}\Big]\mathbf{B}\,\mathrm{dx} \\ = & \int_\Omega H^n(\mathbf{B}\cdot\boldsymbol{\tau})\,\mathrm{ds}, \quad \text{for any } \mathbf{B} \in \mathbf{V}_h. \end{aligned} \tag{15}$$

We note that at each time step n, (15) is a linear system of equations with positive definite coefficient matrix, which can be solved by standard methods. As soon as we know $\mathbf{A}_h^n$ from (15), we substitute it into (14) and solve the nonlinear system of equations to obtain ψ_h^n. In the practical computations we have used Newton's iteration method to solve the nonlinear system of equations (14).

The following theorem is proved concerning the discrete problem (10)-(12).

Theorem 2.1 *Let $\Delta t \le \eta/2$. Then the discrete problem (10)-(12) has a unique solution. Let further Δt be sufficiently small, then under the regularity assumptions*

$$\psi \in C(0,T;\mathcal{H}^2(\Omega)) \cap H^1(0,T;\mathcal{H}^1(\Omega)), \quad \mathbf{A} \in C(0,T;\mathbf{H}^2(\Omega)) \cap H^1(0,T;\mathbf{H}_n^1(\Omega)),$$

one has the error estimate

$$\max_{1\le n\le M}\Big[\|\,\psi^n - \psi_h^n\,\|_{L^2(\Omega)}^2 + \|\,\mathbf{A}^n - \mathbf{A}_h^n\,\|_{L^2(\Omega)}^2\Big] + \sum_{n=1}^{M}\Delta t\|\,\mathbf{A}^n - \mathbf{A}_h^n\,\|_{H^1(\Omega)}^2 \le C(h^2 + \Delta t^2).$$

Moreover, if in addition $\Delta t \le Ch$ for some constant C independent of h and Δt, one has

$$\sum_{n=1}^{M}\Delta t\|\,\psi^n - \psi_h^n\,\|_{H^1(\Omega)}^2 \le C(h^2 + \Delta t^2).$$

Here the constant C is independent of $h, \Delta t$ and M.

NUMERICAL RESULTS

Numerical experiments have been performed on a SGI workstation using the software package "Finite Element Programm Automatic Generator" created by G.-P. Liang.

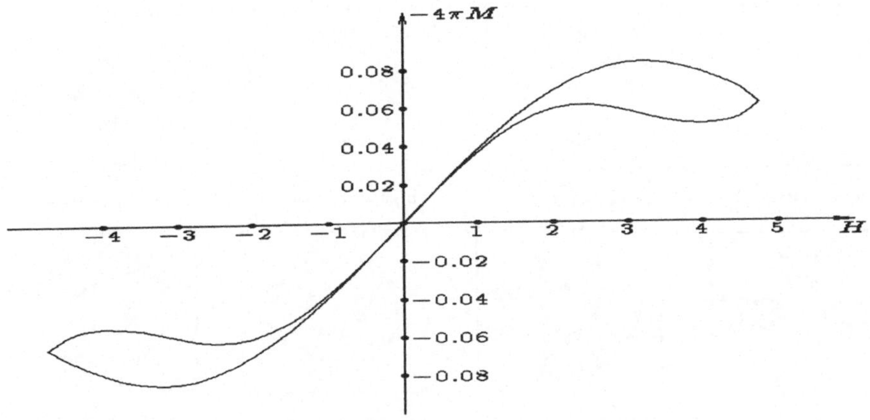

Fig. 1 Hysteresis loop of magnetization against external field for $\kappa = 5$.

In the computations we always take Ω as a rectangle and use piecewise continuous biquadratic polynomials based on a subdivision of Ω into a uniform quadrilateral grid having $N = 1/h$ intervals in each direction. The initial conditions are set to be $\psi_0 = 1/2 + i\sqrt{3}/2$, i.e., a perfect superconducting state.

In the first example we try to compute the hysteresis loop of the magnetization against the external field which indicates that the dynamics of vortices during the superconducting phase transitions depends on the history of the magnetization process. In the computation we take $\Omega = (0,1) \times (0,1)$, $(0,T) = (0,385)$, $\kappa = 5$, $h = 1/40$ and $\Delta t = 0.1$. The external field H is taken to be stepwise function with its increment $|\Delta H| = 0.25$. The field is first increased from zero to 4.75 and then decreased from 4.75 to -4.75, and finally increased again to 4.75. We run the computation at each value of H for a period of 50 steps. The magnetic induction B is calculated by

$$\begin{aligned} B &= \frac{1}{|\Omega|} \int_\Omega \operatorname{curl}\mathbf{A}\ \mathrm{dx} \\ &= \int_0^1 \Big[\mathbf{A}_2(1,x_2) - \mathbf{A}_2(0,x_2)\Big]\ \mathrm{dx}_2 - \int_0^1 \Big[\mathbf{A}_1(x_1,1) - \mathbf{A}_1(x_1,0)\Big]\ \mathrm{dx}_1 \end{aligned}$$

and the magnetization is then $4\pi M = B - H$. The obtained hysteresis loop is depicted in Figure 1.

In the second example we simulate the influence of the external field to the dynamics of vortices. In particular, we are interested in where the vortices start to form, how they interact, and how they disappear. In the simulation we take $\Omega = (0,1) \times (0,1)$, $(0,T) = (0,190)$, $\kappa = 10$, $h = 1/40$ and $\Delta t = 0.1$. The external field H is taken to be stepwise function with its increment $|\Delta H| = 1$. The field is first increased from zero to 9 and then decreased from 9 to zero. We run the computation at each value of H for a period of 100 steps. The contour plots with the minimal and maximal value being zero

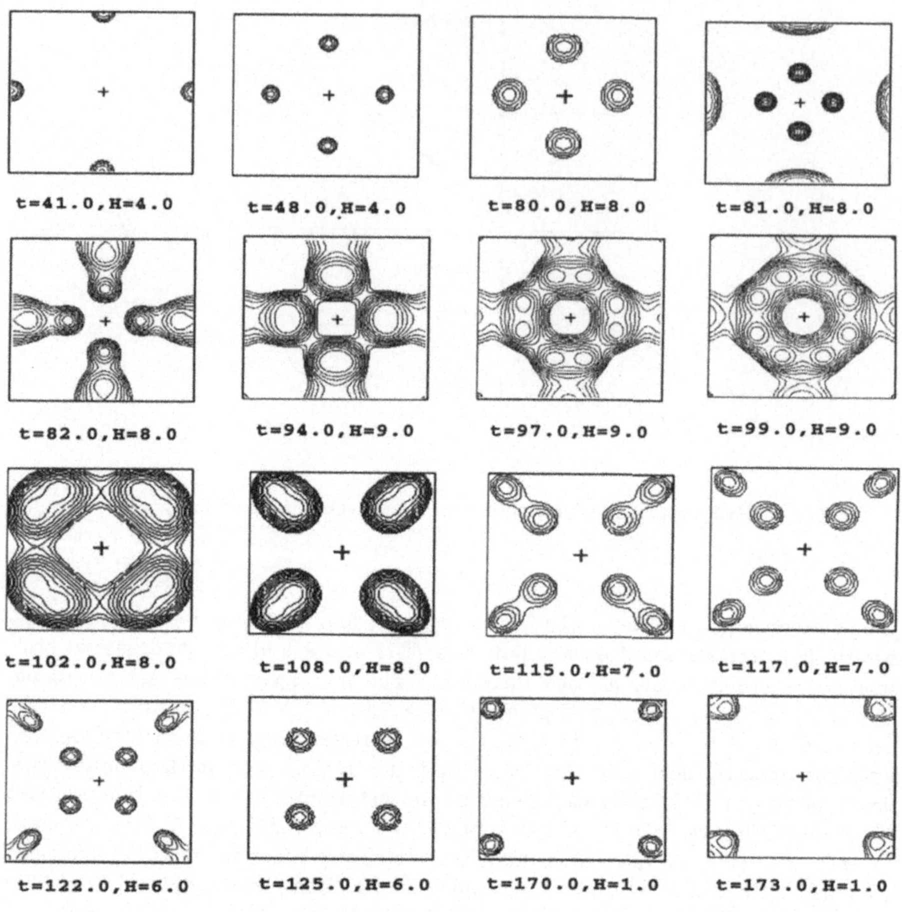

Fig. 2 Contour plots of the magnitude of the density in a symmetric domain.

and 0.1 of the magnitude of the density $|\psi|^2$ are given in Figure 2. We observe that four vortices first start to form near the sides and later four new vortices enter into the domain again from the sides when the field is increased to 8. When the magnetic field is decreased, the eight vortices disappear successively all from the corner.

In the third example we are interested in the dynamics of vortices in a unsymmetric domain. We take $\Omega = (0,4) \times (0,3)$ and $h = 1/80$. The other quantities are taken the same as those in example 2. The contour plots with the minimal and maximal value being zero and 0.1 of the magnitude of the density $|\psi|^2$ are given in Figure 3. We observe that vortices first form on the longer sides of the domain and then they enter into the domain successively from the side. They all disappear from the side when the field is decreased. We also note that the number of the vortices and their unsymmetric distribution are clearly effected by the unsymmetric domain.

In conclusion we remark that, as indicated by the numerical results presented in this paper, the numerical scheme in [2] is effective in simulating the dynamics of vortices in superconductivity. An extension of this algorithm has also been used to simulate the pinning mechanism in the superconducting films having variable thickness in [1].

REFERENCES

[1] Z. Chen, C.M. Elliott and Q. Tang, *Justification of a two dimensional evolutionary Ginzburg-Landau superconductivity model,* to appear.

[2] Z. Chen and K.-H. Hoffmann, *Numerical studies of a non-stationary Ginzburg-Landau model for superconductivity,* to appear in Adv. Math. Sci. Appl.

[3] Z. Chen, K.-H. Hoffmann and J. Liang, *On a non-stationary Ginzburg-Landau superconductivity model,* Math. Meth. Appl. Sci. 16 (1993), 855-875.

[4] Q. Du, *Global existence and uniqueness of solutions of the time-dependent Ginzburg-Landau model for superconductivity,* to appear in Applicable Analysis.

[5] Q. Du, *Finite element methods for the time-dependent Ginzburg-Landau model of superconductivity,* to appear in Comp. Math. & Appl.

[6] Y. Enomoto and R. Kato, *Computer simulation of a two-dimnsional type-II superconductor in a magnetic field,* J. Phys: Condens. Matter 3 (1991), 375-380.

[7] Y. Enomoto and R. Kato, *The magnetization process in Type-II superconducting film,* J. Phys: Condens. Matter 4 (1992), L433-L438.

[8] H. Frahm, S. Ullah and A.T. Dorsey, *Flux dynamics and the growth of superconducting phase,* Phys. Rev. Lett. 66 (1991), 3067-3070.

[9] L.P. Gor'kov and G.M. Éliashberg, *Generalization of the Ginzburg-Landau equations for non-stationary problems in the case of alloys with paramagnetic impurities,* Soviet Phys.-JETP 27 (1968), 328-334.

[10] F. Liu, M. Mondello and N. Goldenfeld, *Kinetics of the superconducting transition,* Phys. Rev. Lett. 66 (1991), 3071-3074.

[11] M. Tinkham, *Introduction to Superconductivity,* McGraw-Hill, New York, 1975.

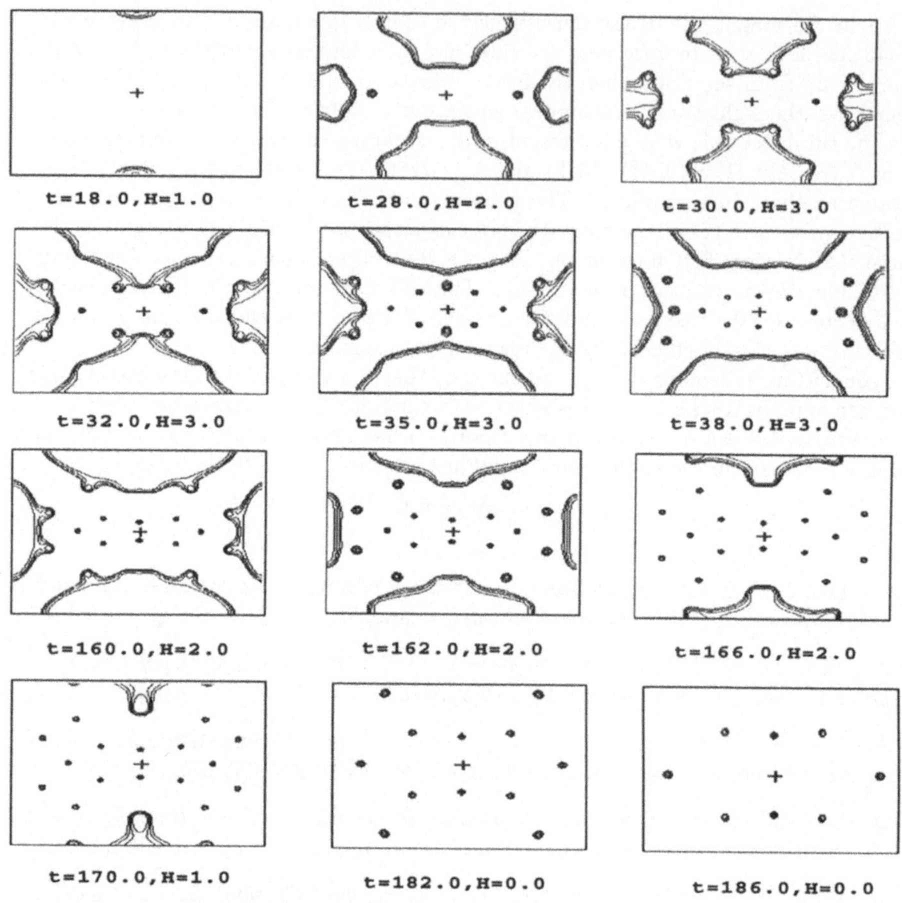

Fig. 3 Contour plots of the magnitude of the density in a unsymmetric domain.

HIGH PERFORMANCE SCIENTIFIC COMPUTING AND ITS APPLICATION IN SOLVING ENGINEERING PROBLEMS

F. Durst
Lehrstuhl für Strömungsmechanik
Universität Erlangen-Nürnberg
Cauerstr. 4, 91058 Erlangen

SUMMARY

Industrial developments that involve fluid flows with heat and mass transfer presently rely heavily on engineering experience gained from experimental investigations and from actual applications. The discipline called High Performance Scientific Computing can provide support to engineering developments through results obtained by numerical simulations. This is most clearly perceptible in the field of numerical fluid mechanics where program developments are underway to compute fluid flows with complex boundaries and involving heat and mass transfer. At the Institute of Fluid Mechanics at the University of Erlangen-Nürnberg, the development of computer programs is underway in cooperation with institutes of mathematics and computer science at the Technical University of Munich and other institutes at the University of Erlangen-Nürnberg. The present paper summarizes the computer program developments and shows that two-dimensional and three-dimensional computer programs are available to compute steady and unsteady fluid flows of compressible and incompressible media. Multigrid solution algorithms are incorporated into the programs and speed up the solution considerably. All programs are parallelized and can be run on computers of different architectures. The present paper shows how programs of this kind might be used in various industrial applications where fluid flow phenomena involving heat and mass transfer are important.

INTRODUCTION

In many fields of industry, gas and liquid flows are employed to intensify or control heat and mass transfer processes. It is the aim of most developers to optimize fluid flow equipment so as to achieve higher levels of heat and mass transfer with less energy consumption. At present, the realization of these optimization efforts is entirely based on experimental investigations that involve measurements performed in models of flow equipment. The data obtained in the models are transferred to the real equipment using transfer rules deduced from similarity theory.

If the flow configurations become more complex and especially if complex heat and mass transfer phenomena occur in the investigations, model investigations result in problems relating to the transfer of model data to the flow equipment employed in industry. In addition, in many situations, complex model constructions for intensive experimental investigations result in high costs and in lengthy investigations. All these

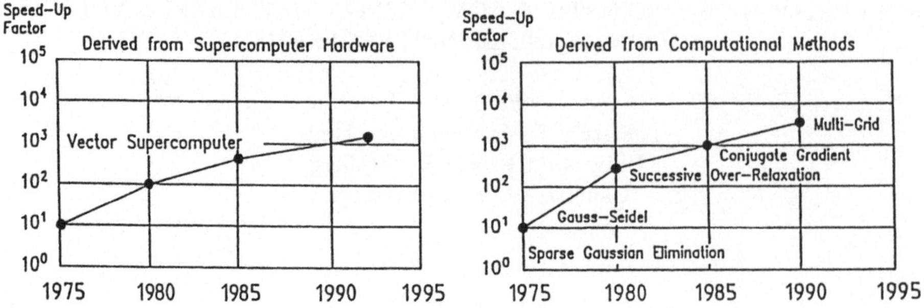

Fig. 1 Developments in the numerical fluid mechanics due to computers and methods

factors consequently result in reduced experimental programs, and so in a reduced number of parameter variations that can be performed, and hence, in incomplete equipment optimization. In most cases, engineers are satisfied with improvements of existing equipment and then sell an attempt of full optimization of the investigated equipment.

There are clear signs that future developments in industry will increasingly rely on numerical simulations to solve optimization problems. This trend is supported by the increased speed with which basic equations of fluid mechanics can be solved by employing modern numerical algorithms to solve the available set of partial differential equations for conservation of mass, momentum, energy, and chemical species (Fig. 1a). In addition, increased computer power has become available over the years allowing more rapid computations on large computers (Fig. 1b). Positive developments, indicated in Fig. 1a and 1b, are continuing. It is foreseeable that they will result in further increases of computer power, hence permitting the direct numerical simulation of complex flow phenomena with heat and mass transfer.

FINITE DIFFERENCE EQUATIONS AND SOLUTION PROCEDURE

The conservation equations solved for fluid flow problems only, result in a set of partial differential equations that can be written in the form of a common transport equation, as given below:

$$\frac{\partial}{\partial x}(\rho U\Phi) + \frac{\partial}{\partial y}(\rho U\Phi) + \frac{\partial}{\partial z}(\rho U\Phi) = \frac{\partial}{\partial x}\left(\Gamma\frac{\partial\Phi}{\partial x}\right) + \frac{\partial}{\partial y}\left(\Gamma\frac{\partial\Phi}{\partial y}\right) + \frac{\partial}{\partial z}\left(\Gamma\frac{\partial\Phi}{\partial z}\right) + S_\Phi + S_{\Phi P}\,.$$

The transport variable ϕ is set to 1 to get the mass conservation equation. The velocity components U, V and W, $U_1 = U$, $U_2 = V$ and $U_3 = W$, as well as the k- and the ϵ-quantities for the turbulence models, result in the source terms given below:

Φ	S_Φ	$S_{\Phi P}$	Γ
1	-	-	-
U	$\frac{\partial}{\partial x}\left(\Gamma\frac{\partial U}{\partial x}\right)+\frac{\partial}{\partial y}\left(\Gamma\frac{\partial V}{\partial x}\right)+\frac{\partial}{\partial z}\left(\Gamma\frac{\partial W}{\partial x}\right)-\frac{\partial p}{\partial x}$	$\overline{S_{P,U}}$	$\mu+\mu_t$
V	$\frac{\partial}{\partial x}\left(\Gamma\frac{\partial U}{\partial y}\right)+\frac{\partial}{\partial y}\left(\Gamma\frac{\partial V}{\partial y}\right)+\frac{\partial}{\partial z}\left(\Gamma\frac{\partial W}{\partial y}\right)-\frac{\partial p}{\partial y}$	$\overline{S_{P,V}}$	$\mu+\mu_t$
W	$\frac{\partial}{\partial x}\left(\Gamma\frac{\partial U}{\partial z}\right)+\frac{\partial}{\partial y}\left(\Gamma\frac{\partial V}{\partial z}\right)+\frac{\partial}{\partial z}\left(\Gamma\frac{\partial W}{\partial z}\right)-\frac{\partial p}{\partial z}$	$\overline{S_{P,W}}$	$\mu+\mu_t$
k	$G_k-\rho\epsilon$	$\overline{S_{P,k}}$	$\mu+\frac{\mu_t}{\sigma_k}$
ϵ	$\frac{\epsilon}{k}(C_1G_k-C_2\rho\epsilon)$	$C_{\epsilon 3}\frac{\epsilon}{k}\overline{S_{P,k}}$	$\mu+\frac{\mu_t}{\sigma_\epsilon}$

$$G_k=\mu_t\left[2\left\{\left(\frac{\partial U}{\partial x}\right)^2+\left(\frac{\partial V}{\partial y}\right)^2+\left(\frac{\partial W}{\partial z}\right)^2\right\}+\left(\frac{\partial U}{\partial y}+\frac{\partial V}{\partial x}\right)^2+\left(\frac{\partial U}{\partial z}+\frac{\partial W}{\partial x}\right)^2+\left(\frac{\partial W}{\partial y}+\frac{\partial V}{\partial z}\right)^2\right]$$

$$\mu_t=C_\mu\rho\frac{k^2}{\epsilon};\ C_\mu=0.09;\ C_1=1.44;\ C_2=1.92;\ \sigma_k=1.0;\ \sigma_\epsilon=1.3;\ C_{\epsilon 3}=1.87$$

For heat and mass transfer problems, the energy equation can also be formulated as a transport equation of the type given above and similar equations for chemical species can be formulated.

In the computer programs FASTEST-2D and FASTEST-3D, the above set of equations is solved using finite volume techniques. All programs incorporate multigrid techniques to speed up the solution procedure. Program versions exist for serial computers and for parallel computers. For special cases, vectorized programs have been set up working also on serial and parallel computer architectures.

EXAMPLES FOR SOLUTIONS OF ENGINEERING PROBLEMS

Many publications which have appeared in recent years, report euphorically about the future possibilities of numerical fluid mechanics and the wide range of applications in engineering. Some papers predict that computational flow investigations will replace experimental fluid flow studies. Replacing experimental techniques by numerical computations is, however, not the primary aim of the efforts within FORTWIHR. It is the aim of the development work to bring out computer codes that permit complementary results to be obtained through numerics in order to shorten development processes which are at present entirely based on experimental methods. The strategy based on this is to be illuminated by some examples briefly described in the following subsections. They constitute only a few examples of the extensive work which is being carried out within FORTWIHR in various fields of fluid dynamics.

Vehicle aerodynamics

The Institute of Fluid Mechanics of the University of Erlangen-Nürnberg (LSTM-Erlangen) has wind tunnels available that permit experimental investigations of ground bound vehicles (cars, trains, bobs, etc.) employing high-tech measuring systems. Fig. 2 shows a Long Range LDA System measuring the velocity field around a car model. Together with pressure distributions and drag and lift force measurements of side force and accompanying moments (6 component balance), it is possible to obtain a complete picture of the influence of the flow field on the car.

Despite the detailed measurements that can be carried out with models in the wind tunnel of the LSTM-Erlangen, the transfer of the results to real ground-bound vehicles is extremely difficult, especially the results measured between the vehicle and the ground. The reason for this is that the wind tunnel investigations of ground-bound vehicles require the ground to move with the velocity of the oncoming air flow. In addition, the wheel-ground interaction should be the same in the model study and the real object. Since the wind tunnel of LSTM-Erlangen is not equipped with a moving belt, the effect of the wall on the flow underneath the vehicle is difficult to obtain experimentally, i.e. the data obtained for the flow underneath the model are not transferable to the flow between the vehicle and the road. This fact has resulted in the development of a strategy for aerodynamic investigations of ground-bound vehicles which employs numerical computations complementary to experimental studies. Experimental investigations are first carried out for the vehicle with fixed ground. Then, a direct comparison is sought for experimental and computer results for fixed wind tunnel ground. If the agreement between measurement and predictions is satisfactory for practical conclusions from the both of the results, numerical computations are repeated for a moving ground floor. Within the computer, it is easily possible to change the boundary conditions in this way and to compute the effect which the motion of the wind tunnel floor will have on the flow field beneath the vehicle. Fig. 3 shows a vehicle for which laser Doppler measurements have been carried out and compared with numerical predictions. The agreement reached between the experimental and numerical data is satisfactory for most practical conclusions drawn from the investigations. Hence, the computation that a motion of the ground floor will yield a change in sign of the lift force enforced to the back of the vehicle can be taken as guaranteed for the corresponding large vehicle moving on the road.

Aerodynamic of Airfoils and Wings

A pronounced need exists for more sophisticated methods for scale-up and wind tunnel wall corrections in experimental aerodynamics of airfoils and wings in order to take into account the pressure distributions enforced by the wind tunnel geometry. This is especially true for tests of high-lift and VTOL (Vertical Take-Off and Landing) configurations where, in addition to the geometrical blockage the aerodynamic blockage becomes equally important. The effects due to aerodynamic blockage are fundamentally different from those of geometrical blockage and cannot be accounted for in a similar way. There have been many attempts in the past to reduce these problems, using advanced numerical simulation techniques coupled with high performance computers.

Another area in experimental aerodynamic research where the scale-up of wind tunnel results is not completely understood is the field of laminar to turbulent transition

Fig. 2 LDA-system at the LSTM-Erlangen wind tunnel and its implementation to the aerodynamical investigations of ground-bound vehicles

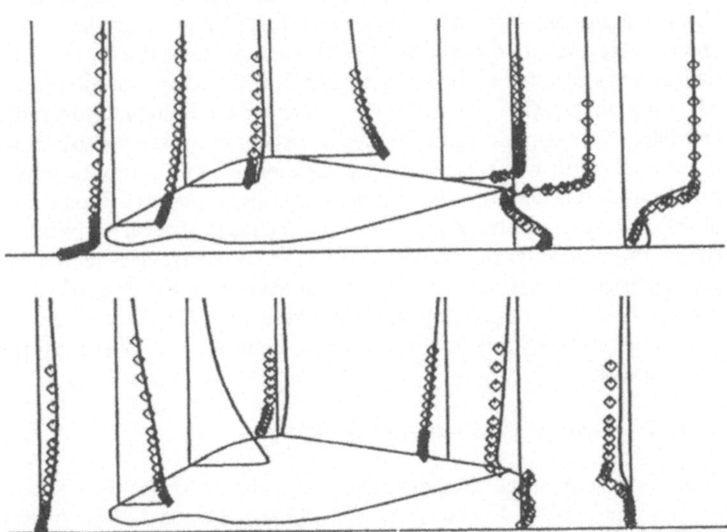

Fig. 3 A comparison of numerical calculations with LDA measurements for a model vehicle near the tunnel ground

Fig. 4 Semi-conductor Laser-Doppler-Anemometer during flight tests

or more specifically in the development of laminar flow wings for aircraft. Recent research efforts have shown that progress still depends largely on flight testing, because full or nearly full scale Reynolds numbers are mandatory to obtain relevant results, and the intensity and scales of wind tunnel turbulence always influence the laminar flow and the transition to turbulence. Flight testing on aircraft is extremely expensive and there is still a lack of advanced measuring techniques for in-flight measurements. Fig. 4 shows a flight experiment using a laser Doppler anemometer consisting of a semi-conductor laser and photodetector. With such an instrument, in-flight data can be obtained and the results can be compared with wind tunnel investigations using an airfoil of the same shape. Comparing computations for the in-flight and wind tunnel test conditions with the corresponding experimental results permits conclusions to be drawn regarding the applicability of computer programs for scale-up computations. Boundary layer calculations using low-Reynolds-models have been performed and indicate good agreement with measurements (Fig. 5). In these calculations, an empirical correlation is adopted for the determination of separation. There is strong evidence that using a more physically based transition criterion, e.g. derived from the e^n-method, will further improve the prediction of transition positions. These numerical simulations of boundary layers can then serve as a basis for scaling-up procedures, i.e. for the interpretation of experimental results.

Simulation of transonic and supersonic flows

In recent years, numerical methods and codes for the simulation of transonic and supersonic flows have been developed at the LSTM-Erlangen. These numerical codes solve the compressible Euler and Navier-Stokes equations in conservation law form for laminar and turbulent subsonic, transonic, and supersonic flows. The development of the codes is based on state-of-the-art finite volume methods as well as on parallelization procedures that have been introduced.

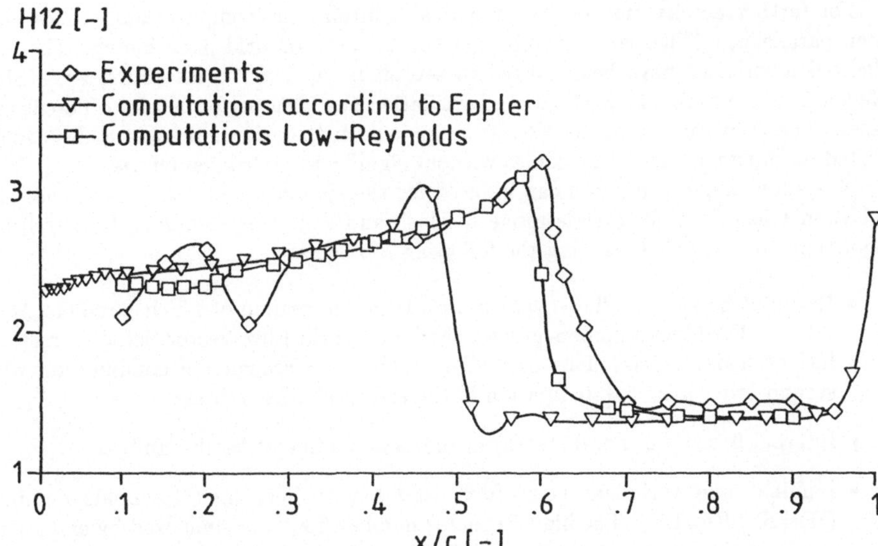

Fig. 5 Shape parameter H_{12} on suction side of E580 airfoil

In more detail, different shock capturing discretization schemes have been developed at LSTM-Erlangen for several transonic and supersonic flows. These schemes are classified in the following categories:

- Flux Vector Splitting (FVS) methods:
  - van Leer FVS
  - modified Steger-Warming FVS
- Riemann solver
- High resolution schemes (Monotone Upstream Scheme for Conservation Law) with second order of accuracy and a five point scheme with up to third order of accuracy.

In the recent years, these methods have been validated by comparisons with experimental results. Furthermore, an implicit algorithm, developed for the time integration, has been coupled with the above numerical methods. The numerical algorithm is a very efficient unfactored implicit method which allows the use of large time steps. For the acceleration of the convergence, different numerical methods and techniques have been developed. These are:

- the local time step
- the mesh sequencing
- the local solution.

For further acceleration of the numerical solution, the computational codes have been parallelized. The parallelization method is based on grid partitioning. The parallelized algorithms have been ported to several parallel platforms (Parsytec, KSR1, Meiko, Convex) available at LSTM and validation has been performed for different test cases. The employed parallelization strategy is such that the numerical codes can be ported on different parallel machines without significant additional efforts.

The above numerical codes can be used for the simulation of two dimensional and axisymmetrical flows for a wide range of Mach and Reynolds numbers. In detail, the algorithms are capable of solving the following problems:

- External flows (e.g. flow over an airfoil) in the regime of Mach numbers $M = 0.1 - 2$. For Mach numbers greater than 2 a hybrid Flux Vector Splitting method has been developed. This method uses the FVS schemes in combination with second order artificial dissipation in the strong shocks regions
- Internal flows (e.g. nozzle flows) in the same regime of Mach numbers
- Laminar and turbulent flows from very low to very high Reynolds numbers ($Re=10^2$, $Re=10^7$). The high Reynolds number flows are simulated by employing the Baldwin-Lomax algebraic turbulence model, widely used in the aerodynamic community
- Problems with variable boundary conditions (adiabatic and non-adiabatic walls, free stream conditions etc.)
- Shock wave/boundary layer interaction
- Separated flows and flows with a high angle of incidence.

The extension of the above numerical capabilities in 3D transonic and supersonic flows has been started and programs should be available in the near future.

Chemical Vapor Deposition

Metalorganic chemical vapor deposition (MOCVD) has become an established technique for growing thin, high purity, epitaxial films of compound semiconductors for modern optoelectronic applications like solar cells, lasers and high electron mobility transistors. Numerous MOCVD reactors and processes have been successfully developed for experimental and industrial purposes. The development of MOCVD techniques has been based largely on empirical findings and step-by-step improvements by trial and error procedure; therefore, the optimization of equipment could be greatly enhanced by use of simulation. In recent years, considerable progress has been made in the mathematical modelling of MOCVD processes. The objective of the various mathematical models employed in numerical computations is to relate performance measures (growth rate, uniformity, composition, and interface abruptness) to operating conditions (pressure, temperatures, reactant concentrations etc.) and reactor geometry. Besides the practical advantages of gaining numerical information on the growth rate of films, the models provide insight into the basic MOCVD physicochemical processes that are made up of interactive transport processes and chemical reactions. A model for epitaxial growth in horizontal and vertical MOCVD reactors has been developed at

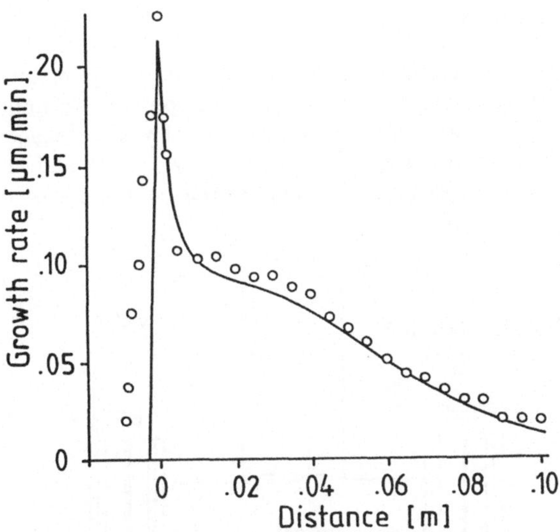

Fig. 6 Growth rate. The horizontal tilted susceptor reactor. Comparison of experimental results (circles) with model predictions (solid line)

the LSTM during the last two years. The MOCVD modelling can be subdivided into three areas: flow prediction, mass transfer prediction, and chemical reactions. The flow in MOCVD reactors can be satisfactorily predicted by taking into account the detailed boundary conditions including the processes of heat conduction in the reactor walls, external and internal radiation between susceptor and walls, external radiation, natural or forced convective heat transfer to the ambient, and three-dimensional effects whenever the various partial processes contribute to the formation of deposited films. The modelling of the mass transfer includes the convective, diffusive and thermodiffusive transport of all active chemical species as well as homogeneous (gas-phase) and heterogeneous (surface) reactions. The described model has been implemented in a multigrid finite volume solution procedure for flow, heat- and mass transfer calculations (plane and axisymmetrical) including the view factors calculations code. Due to the multigrid solution procedure, computing times were reduced by an order of magnitude or more in comparison with standard methods ordinarily used for MOCVD modelling. The model is applied to MOCVD of GaAs from TMGa and AsH_3 under widely varying conditions. Results of the calculations are compared with experimental growth data. The excellent agreement of this model with experimental data is observed in Fig. 6, which depicts the growth rate of GaAs along the susceptor. After the agreement is achieved, the model can be used for process and equipment optimization purposes in order to get desirable growth conditions in the reactor.

The Czochralski process

The production of single crystalline materials like silicon, germanium and gallium-arsenid by crystal growth from the melt is a very important technique for the semiconductor industry. The quartz crucible, which is filled with a semiconductor melt, is located in a block of graphite (Fig. 7). In addition to quartz and graphite, the crucible may also be manufactured by aluminiumnitride, aluminiumoxide and pyrolytic boronnitride.

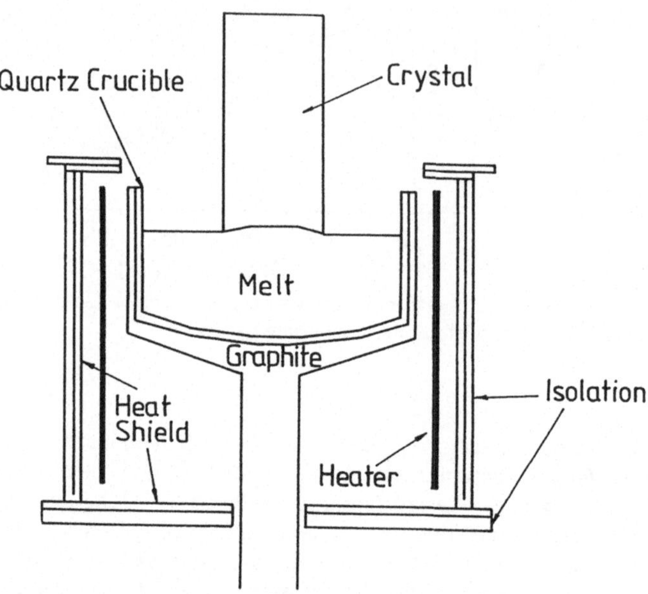

Fig. 7 Schematic Czochralski crystal growth equipment

The growing crystal is pulled out of the melt with the growth velocity of about $3-15\ \frac{cm}{h}$. During the growth doping materials are incorporated into the crystal lattice. These doping materials should be spread over the bulk crystal very regularly in order to provide homogenous physical properties.

It is also very important to avoid dislocations inside the crystal lattice. Hence, the growing crystal must meet an orientation of growth, which is preset by a small seed crystal. At the beginning of the growth process the crystal is pulled out in such a way that its diameter increases until the scheduled diameter is reached. Then, the diameter of the crystal must be kept constant by controlling all process parameters.

An important impact on the growth process is given by the thermal boundary conditions. In order to get axisymmetrical thermal boundary conditions, it was found advantageous to rotate the crucible and the crystal. The global thermal conditions in a crystal puller are linked to the basic mechanisms of heat transfer. Besides convection and diffusion, the heat radiation takes a non-negligible part of heat transfer since the mean temperature inside a crystal puller is approximately 1700 K.

The Czochralski process is connected with many technological questions, which hitherto could not be solved completely. This is due to the complex construction of the

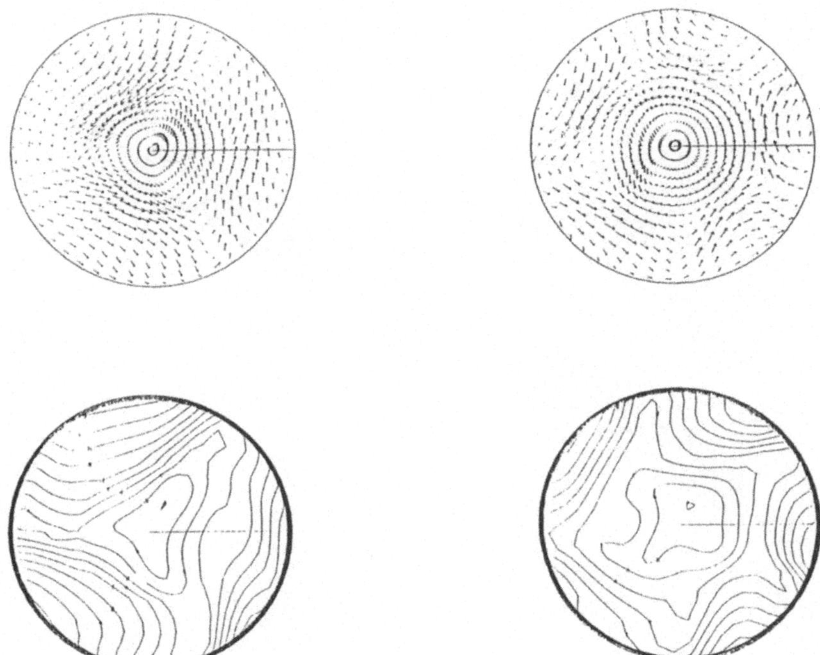

Fig. 8 An example for a non-axisymmetrical flow pattern in a silicon melt: the picture shows a three(five)-folded wave pattern (isotherms) which develops under the impact of both buoyancy and Coriolis forces. Left: $\omega_{crucible} = 3\ rpm$, Right: $\omega_{crucible} = 5\ rpm$

crystal puller and the lack of knowledge about heat transfer, flow behaviour, and instabilities in the melt. The flow in the melt extensively impacts the growth behaviour of the crystal and the incorporation of doping materials into the crystal lattice. Unfortunately, the melt flow can not be investigated by direct visualization methods or Laser-Doppler anemometry, since the melt is not transparent for light rays to pass.

Very recent flow visualization experiments using ultrasonic waves and X-rays seem to be more promising to get very detailed information about the melt flow. Applying these new techniques, flow instabilities were investigated and it was shown that in most cases the flow is non-axisymmetrical and unsteady. A typical example of a non-axisymmetrical flow is shown in Fig. 8. These wave patterns are responsible for low frequency temperature fluctuations, which tremendously impact the growth of the crystal.

From this point of view, three-dimensional, time-dependent numerical simulations of the melt flow seem to be a very promising approach in order to get more detailed information about flow instabilities and the dependency of heat transfer on the flow pattern. In future, flow behaviour should also be numerically investigated in the turbulent regime, since turbulence is likely to occur in real industrial Czochralski pullers.

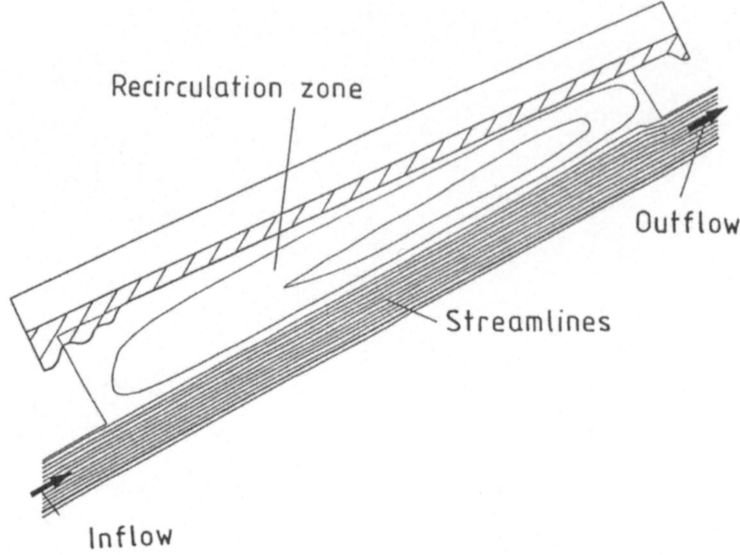

Fig. 9 Streamlines in the outer ventilation channel of a tiled roof

Ventilation of Roof Constructions

Computer codes available at LSTM-Erlangen can also be employed to support developments in small and medium size companies, i.e. their application is not limited to high tech problems usually of concern to larger companies, as indicated in the above subsections. To demonstrate this, various applications have been set up to demonstrate the employment of FASTEST-2D and FASTEST-3D to engineering problems typical for small and medium size companies. One of the examples chosen for the presentation in this paper, relates to the question of whether ventilations of roof constructions are needed to prevent humidity damage to the wood used on the tiles of the roof. The application of FASTEST-2D to solve the momentum, heat and mass transfer problems in the inner and outer subchannels of the roof construction resulted in confirmation of experience laid down in the so called "Dachdeckerregeln". Computational results proved that ventilation of roof construction is a very effective means of removing the humidity contents entering the roof construction through the insulating material.

In Fig. 9, computed velocities under a tile are reported. Similar results are available for the distribution of humidity concentration.

CONCLUSIONS AND FINAL REMARKS

In this paper, examples are shown that indicate how High Performance Scientific Computing can be used to supplement information obtained from experimental studies of flows in engineering applications. The presented examples have indicated that it should not be the aim of numerical fluid mechanics using advanced methods of solving the basic equations of fluid mechanics to replace experiments but rather to support experimental results. In this way, more advanced developments are feasible permitting more rapid developments and therefore shorter development times. The development work carried out in section 1 of FORTWIHR aims at the provision of computer codes for flow predictions to be applicable in engineering development work.

COMPUTATIONAL FLUID DYNAMICS WITH FIRE

ON MASSIVE PARALLEL COMPUTERS

H. Fischer, C. Troger

BMW-AG, Entwicklung Antrieb

80788 München

SUMMARY

At BMW the main production code for 3-dimensional fluid flow simulations is currently ported to a massive parallel processing (MPP) computer system with distributed memory. Special emphasis has been taken to integrate the MPP system seamlessly in the productive computer aided engineering (CAE) environment at BMW. Several decomposition algorithms and hidden communication have been introduced to minimize overhead compared to serial computers. Production jobs using already ported functionalities of the code are submitted to the MPP via nQS (nCUBE Queueing System) in a multiuser batch environment

INTRODUCTION

The use of computer simulation of physical processes is widley seen as a means of meeting todays developement constraints such as cost reduction, improved product quality and shorter developement cycles. For a succesful contribution to the final product, simulation has to start before the first hardware is build and has to accompany the whole design phase. A close collaboration between simulation, design and test beds has to be established. The simulation cycle itself has to be optimized for speed and efficiency to be able to meet tough production time scales. Therefore continuous improvements have to be made on the main constituents of a simulation cycle: Interface to Computer aided design (CAD), mesh generation, calculation and visualisation of the results. Soft- and hardware deserve equivalent attention. Looking at the actual flow simulation, a sound concept is needed to match the steady increasing demands for calculational power. MPP with distributed memory seems to be the natural way to provide nearly unlimited extendable computing power for the future, if the production software is available on such platforms.
In the following a presentation of the computational fluid dynamics (CFD) code FIRE together with its use at BMW will be given. After this the motivations and expectations when choosing the MPP platform will be explained and selected technical topics like porting strategies, decomposition and communication algorithms will be discussed in some detail. Some benchmark results and a glance into the future will make the conclusion

FIRE - FLOW IN RECIPROCATING ENGINES

FIRE is an integrated software package comprising CAD interfaces, pre- and postprocessing and a calculation module. The basis of the calculation program are the conservation equations for mass, momentum and energy in three dimensions. The flow can be compressible or incompressible, steady or unsteady, laminar or turbulent. The basic methods used are:

- Finite volumes in nonorthogonal multi-block structured or unstructured grids
- Moving grid and rezone (change of grid topology during the calculation)
- Discretisation schemes of different order
- Higher order turbulence models (k-eps., RNG, RSM)
- A posteriori error estimation
- Interfaces for user-supplied program modules

- Gauß and conjugate gradient equation solvers
- SIMPLE solution algorithm

A variety of physical submodels necessary for the simulation of flow in internal combustion engines are available:

- Spray (Lagrange droplet model)
- Wall film (Couette flow)
- Two phase flow (Interacting continua)
- Combustion (Arrhenius-, turbulent mixing- and pdf-models)
- Conjugate heat transfer
- Porosities
- Linear acoustics

FIRE has an interface to a one-dimensional gas dynamics calculation program. This allows to calculate the dynamics of the complete engine including intake and exhaust manifolds and resolving simultaneously selected regions (e.g. the combustion chamber) with the 3-dimensional method.

FIRE - HARDWARE AND APPLICATIONS AT BMW

The design of FIRE supports the distributed use of its modules in a heterogeneous UNIX environment. Platforms at BMW include SUN, SGI, IBM RS600 workstations and CRAY-YMP and nCUBE2s supercomputors. The FIRE-manager (FIMAN) allows to use FIRE transparently in the computer network at BMW on distributed resources like workstations, CPU servers, File servers and DISPLAY servers. Fig 1 gives a sketch of the CAE computer environment at BMW.

Simulation is used in the engine development for optimising components and as a tool for analysing physical processes which determine the global behaviour of the engine. Main application areas include calculation of flow in intake and exhaust manifolds [1], mixture formation, in-cylinder flow, combustion [2] and water cooling jackets. FIRE is also used in the car-body developement for the calculation of flow in passenger compartements and air-conditioning systems. At BMW-Rolls-Royce (BRR) flow and mixing in jet-engines is calculated with FIRE.

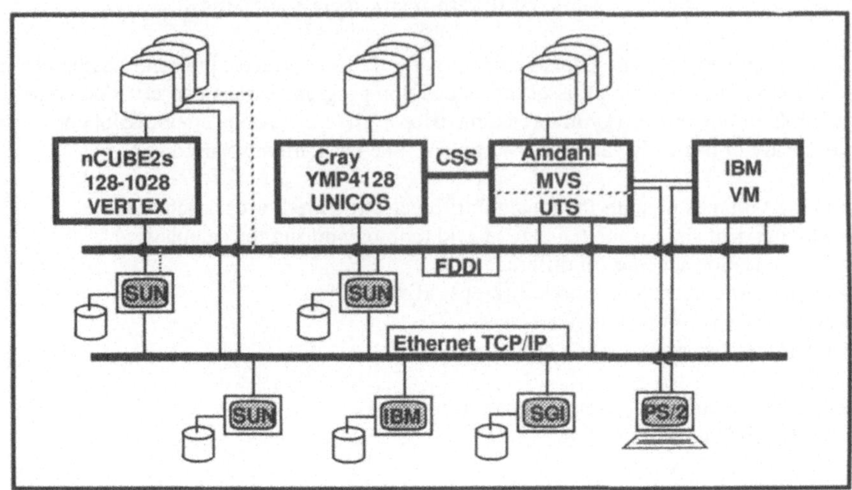

Fig. 1: CAE Computer Environment at BMW

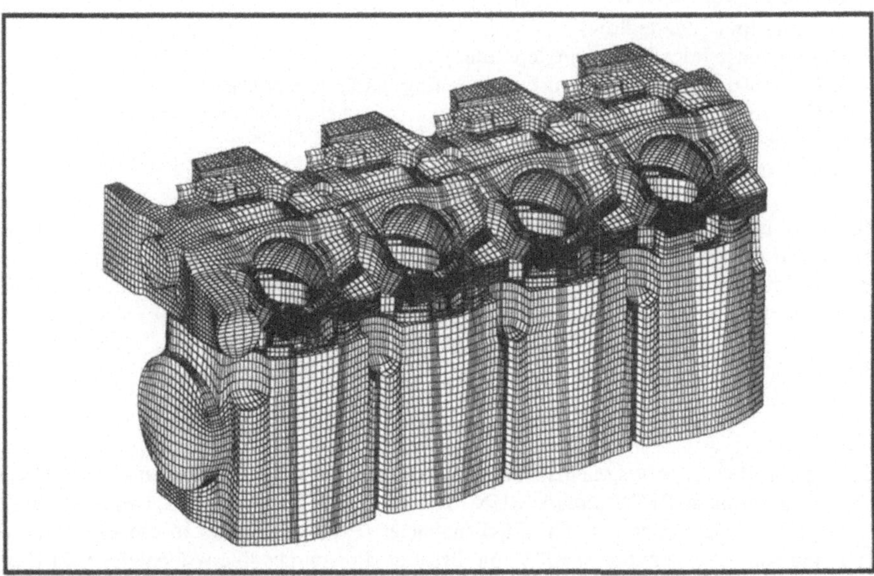
Fig.2: Water Cooling Jacket of the 2-Valve 4-Cylinder Engine; Computational Mesh

Figure 2 shows as an example the computational mesh of a water cooling jacket of the 2-valve 4-cylinder engine. In figure 3 contour values of the Nusselt number as a result of a steady state calculation are drawn. They give the engineer the desired information about the location of convection enhanced heat transfer areas.

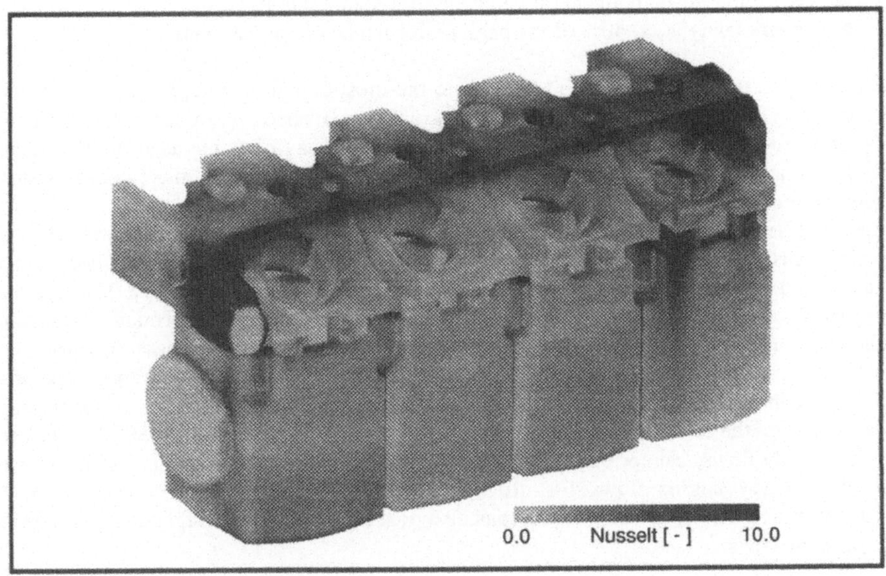

Fig. 3: Water-Cooling Jacket of the 2-Valve 4-Cylinder Engine; Levels of Calculated Nusselt Number

MOTIVATIONS FOR MPP

If simulation methods want to face successfully the challange of contributing to the quality and time-to-market of advanced products, it is necessary to deliver results:

- in time
- for real world applications
- with a high degree of accuracy

Satisfying this requirements will result in:

- shorter time-scales for each step of the simulation cycle
- complex physical sub-models
- increased problemsize

Looking at the steadily increasing number of users which want to benefit from simulation we will need improved software and extendable harwdare performance to achieve our goal of maturing from research to industrial developement.
Thourough investigations have been made by the authors concerning the performance of the flow-calculation module of FIRE on CRAY vectorcomputers. The main findings have been:

- Turnaround times in a general batch environment are high.
- The efficiency is ca. 40% of the peak performance of the hardware.

The first is due to the fact that CFD belongs to the most demanding applications in terms of memory and CPU time. Such applications are usually „punished" in a general batch environment with very low priorities. Therefore turnaround times may be as big as 10 to 20 times the actually needed CPU time. This results in much too long time spans to make contributions to serious development work.
The second has its roots mainly in the indirect addressing scheme, which is needed for complex 3-dimensional geometries. Two memory fetches are necessary for each data item: one for the index and one for the data itself, and the data is scattered over the memory. Therfore load/ store operations and bank conflicts become a bottleneck and the performance is memory bound. This situation becomes worse when more CPU's are added to the shared memory.
The conclusion was, that the price/performance ratio of convevtional supercomputers is far too high to stay in business for serious contributions to the product developement. Todays super workstations on the other hand, though very appealing at the lower end (i.e. 250000 cells problems for steady flows) cannot offer a perspective at the high end problem classes (i.e. calculation of unsteady engine flows for different loads, injection parameters, etc.). The I/O performance and disk space of workstations also poses constraints for calculation of unsteady and transient flows.

THE MPP PROJECT AT BMW - STRATEGY AND CONCEPTS

After a two years phase of benchmarking and evaluation it was clear that MIMD based MPP's can be a basis for meeting todays and future demands for computing performance. Main features include:

- easy upgradability
- cheap components which result in low overall price
- distributed memory without bottlenecks and bank conflicts
- parallel I/O

A project was started together with BMW, AVL-Graz (the software vendor) and nCUBE (the hardware vendor) to generate a parallel, message passing version of FIRE [3]. nCUBE was choosen as the platform for the port because of its balanced hardware (communication to calculation performance), its parallel I/O subsytem and its availability on the market.
The parallelisation of FIRE had to be a sound investment for the future and should fit for all emerging hardware concepts. To ensure this a strategy was formulated:

- The parallelisation has to be hardware independent as far as possible
- There have to be clearly defined interfaces to hardware dependent features/optimisations
- There is only one basic code. The parallel version is produced out of the sequential one with inserted directives and a pre-compiler
- Parallelisation will proceed in terms of FIRE functionalities
- Several decomposition algorithms have to be implemented
- Decomposition and communication have to be setup on the MPP as far as possible

A sketchy picture of the parallelisation concept of FIRE is given in figure 4. Emphasis was laid from the beginning not only on computing performance, data distribution and comunication but also on parallel I/O with striped file sytems. nCUBE was able to deliver this features in a transparent way for the programmer. Although nCUBE uses a hypercube, FIRE is not restricted to any special communication topology.

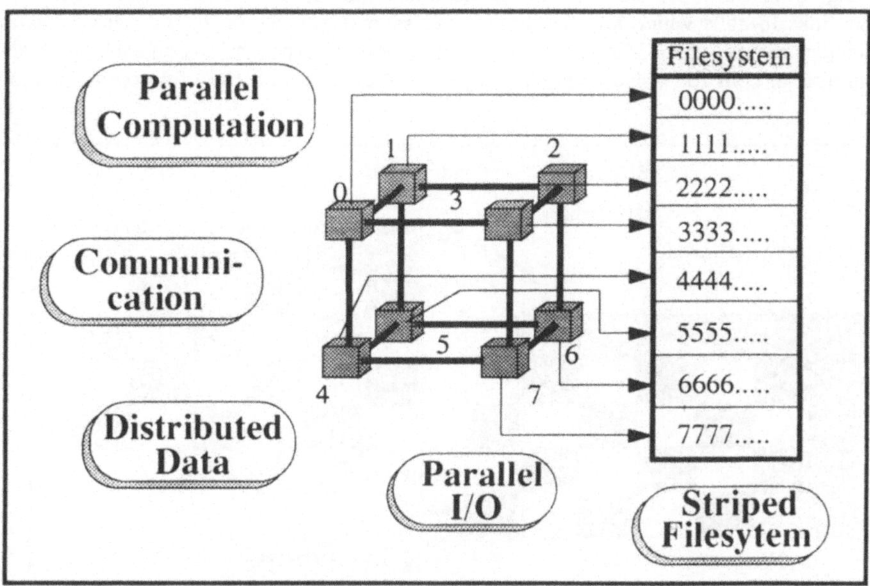

Fig. 4: Parallelisation Concept for FIRE

On a computer with distributed memory only a fraction of the total problem is running on each of the processors, and each processor holds only a small part of the total data in his local adressing space. The equation solver and other tasks in the solution process operating on a certain location of the mesh need values from the neighbour cells, which at interprocessor boundaries (IPB's) may be located on a neighbour processor. Therefore before solving a problem on such a computer two complementary tasks have to be done: to split (decompose) the problem logically and to setup a communication scheme such that all processors know where to get the data at the IPB's from.

COMMUNICATION

Data has to be communicated between processors at certain stages during the computation to distribute/gather I/O data which is the same for all processors, to exchange field values in the solver (Gauss Seidel and Conjugate Gradient) and to exchange values like weighting factors and derived quantities at IPB's to ensure numerical „continuity" between processors.
Figure 5 shows the situation at an IPB in detail, together with the red/black colouring of computational cells for the Gauss-Seidel solver. IPB's are build of an additional external cell layer

around all cells which become boundary cells after splitting the geometry. Values which are imported from the neighbourprocessors are stored in this locations. The processor itself has to export the values in its first inner cell layer into the external cell layer of its neighbours. If two domains are put together they would overlapp with their corresponding external and first internal cell layers (see fig. 5).

Depending on the numerical scheme and the used decomposition technique, a cell at an IPB can have links to cells which are distributed over several processors. In this case it has to export its data several times. The ratio of the total number of export events on each processor to the number of cells for which computation has to be done is a measure for communication intensity.

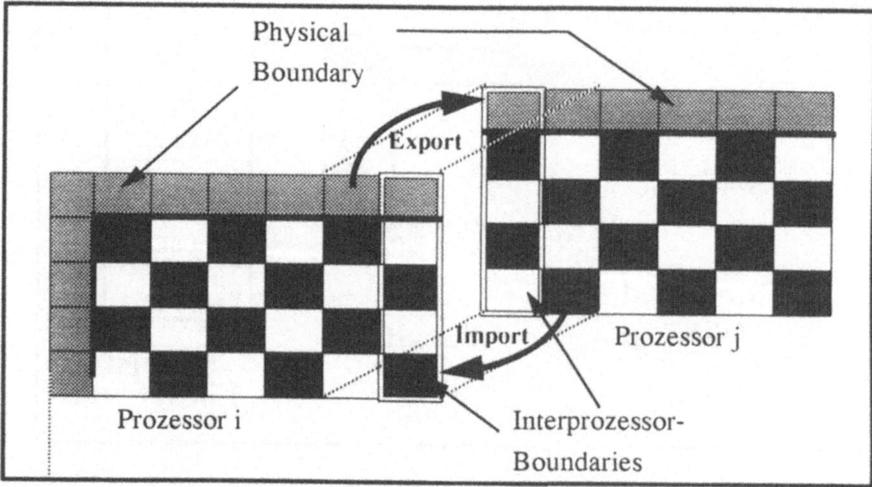

Fig. 5: Communication at Interprocessor Boundaries

For the Gauss-Seidel solver a scheme has been developed to hide the communication behind the computation using nCUBE's asynchroneous message passing libraries. This scheme is shown in fig. 6.

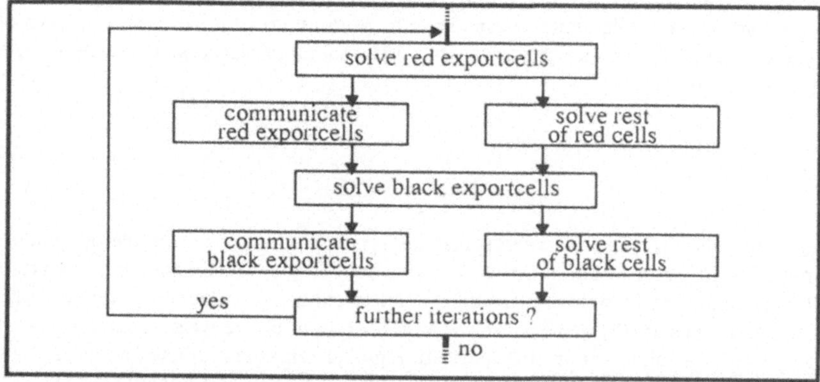

Fig 6: Hidden Communication in Red/Black Ordered Gauss-Seidel Solver

DECOMPOSITION ALGORITHMS

Figure 7 shows on a small example the four decomposition algorithms included in the FIRE preprocessor, namly the data decomposition and three different types of domain decompositions. Each part of the mesh for which calculation is done on a different processor is shown in a different greyscale. The boundaries between this areas constitute the IPB's and vary in size and shape depending on the type of decomposition.

The simplest variant of these is the data decomposition (DD). The arrays on which the relevant data is stored are cut in as many parts as processors are available. Due to the arbritray order of the indirectly addressed data on the arrays, locations which lie geometrically in a certain neighbourhood may not reside on the same processor but can be scattered over a number of processors which must not be direct neighbours in the communication network of th MPP.

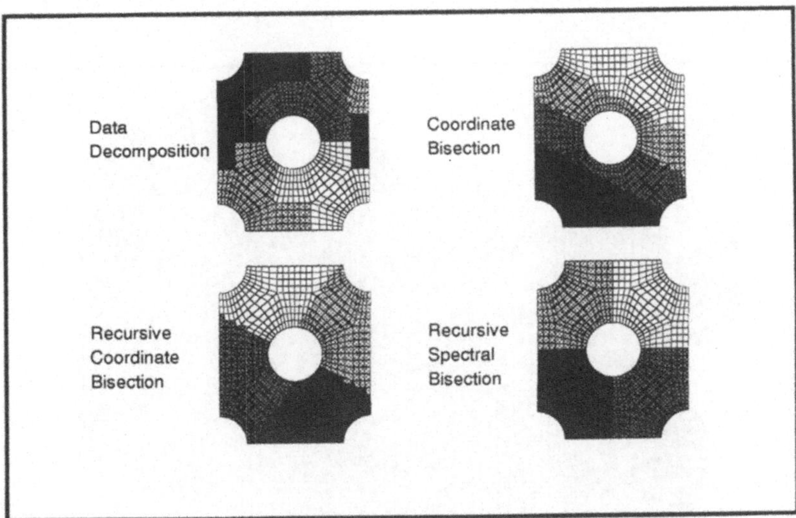

Fig. 7: Decomposition Techniques on an Unstructured Grid

Domain bisection algorithms on the other hand are designed in a way that all cells in a certain geometric neighbourhood are mapped onto the same processor. The intention is to keep the communication intensity at a low value.

The coordinate bisection (CB) accomplishes this task directly in geometric space. The computational domain is searched for its largest elongation and all cells are ordered according their position along this direction. In the next step cell-numbers are assigned to processor. This can be done in two ways: Either by direct partitioning of the ordered array in as many parts as processors ar available or by division in two parts and applying the whole procedure recursively on each resulting sub-domain (recursive coordinate bisection, RCB).

Recursive spectral bisection (RSB) on the other hand utilizes a synthesis of graph theory and eigenvalue problems. The topological structure (graph) of the mesh is mapped on a matrix A in a way that for

$i \neq j$:	$A(i,j) = 1$ if i and j have a logical link in the mesh
	$A(i,j) = 0$ otherwise
$i = j$:	$A(i,j) =$ minus the total number of logical links for this element

The eigenvalues and eigenvectors of this Laplacian matrix are computed and the components of the second eigenvector (Fiedler vector) contain some distance information for the mesh elements [4]. The grid cells are then ordered acoording to the size of their corresponding components in the fiedler vector and divided into two parts. The procedure is applied recursively to each resulting sub-domain until all cells are distributed over the available processors.
In figures 8 to11 the results are shown of applying all four decomposition algorithms to the water cooling jacket of fig. 2 for 128 processors. In looking at this pictures, one has to keep in mind that only the surface part of the decomposition is visible. A judgement of the quality of the decomposition in terms of number of communication cells and number of neighbours (which finally accounts for the communication expense during the computation) can not be made on the basis of this pictures alone. A more quantitative analysis is given later on. Nevertheless the characteristic features of each type of decomposition can be grasped in an intuitive way from this figures.

Fig. 8: Water Cooling Jacket of the 2-Valve 4-Cylinder Engine; Data Decomposition for 128 Processors

The mesh for the water cooling jacket is topologically a block with holes which yields a very regular numbering of the cells. So the data decomposition operating on the array indices results in a quite regular pattern of the subdomains. The geometry is cut into slices (fig.8). The drawback of this approach is that the value of the communication intensity is high.
The situation is even worse for the coordinate bisection (fig. 9). Here the hexahedral cells used as finite volumes in FIRE lead to very rugged IPB's. The coordinate bisection algorithm cuts the mesh block into skewed slices and the result is a surface with many little steps. This increases the amount of communication relative to the computation, because data has to be exchanged across each face of a cell lying in an IPB.

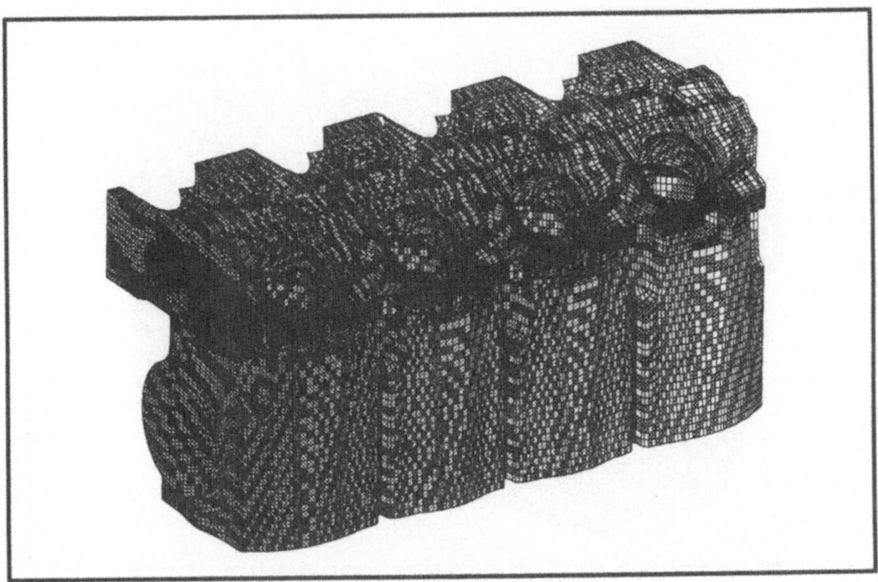

Fig. 9: Water Cooling Jacket of the 2-Valve 4-Cylinder Engine; Coordinate Bisection for 128 Processors

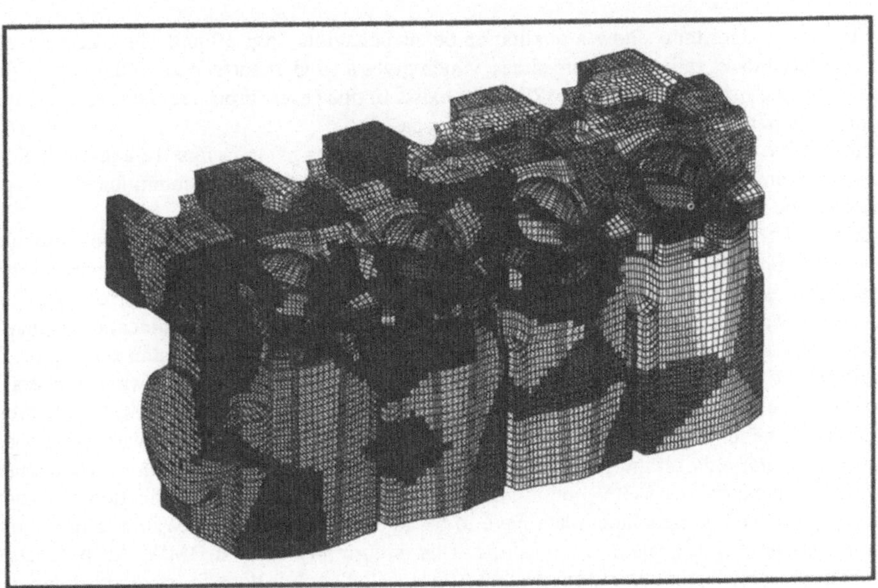

Fig. 10: Water Cooling Jacket of the 2-Valve 4-Cylinder Engine; Recursive Coordinate Bisection for 128 Processors

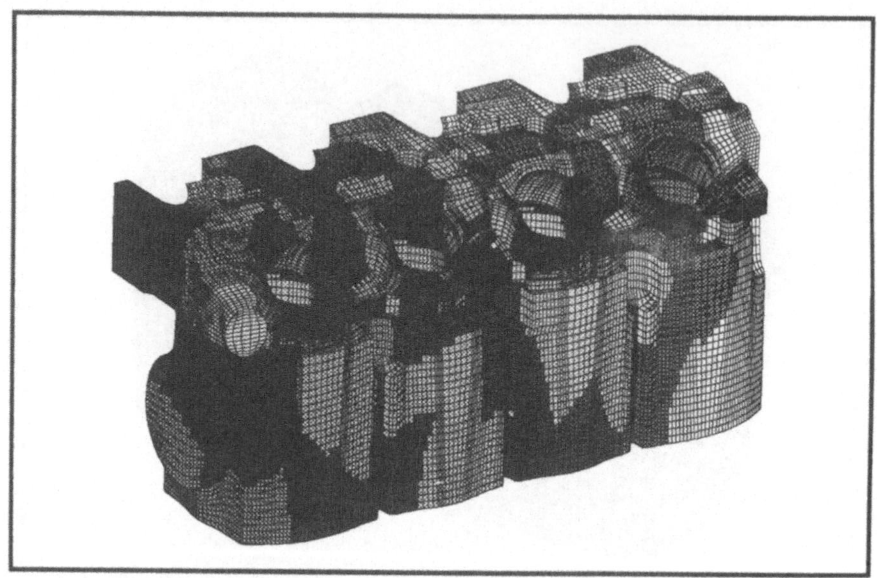

Fig. 11: Water Cooling Jacket of the 2-Valve 4-Cylinder Engine; Recursive Spectral Bisection for 128 Processors

The recursive algorithms show a qualitative better behaviour (figs.10 and 11): The cells are grouped into blocks rather than into slices, which gives a smaller surface to volume ratio. The problem of the rugged shape of the IPB's still exists in this cases. From the pictures alone it is not possible to make a decision between this two cases.
A quantitative comparison of the different decompositions is given in figs. 12 and 13. Fig. 12 shows the communication intensity. A small value corresponds to little communication overhead over the actual computation time.
Looking at fig. 12 the qualitative pictures of figs. 8-11 are confirmed. The coordinate bisection is the worst one. For processor 20 e.g. the number of cells on other processors to which it has to send data is four times larger then the number of cells for which it has to perform calculation. We now see also that the RSB is superior to the RCB in terms of communication intensity. In fig. 13 the number of neighbours with which each processor has to communicate is plotted. In this picture the data decomposition and the coordinate bisection have the lowest numbers of neighbour processors. Therefore the question arises whether communicating less data to more processors is better than communicating more data to less neighbours. The answer to this question depends not only on the geometry of the problem but also on specific hardware features like communication bandwith, communication startup time and communication topology. Some experience and benchmarkdata have to be gathered, before it is possible to draw final conclusions or derive general relationships. This is currently done at BMW. All production computations are benchmarked and, when possible, calculation is done for several decompositions. The results will be published in a forthcoming paper [5].

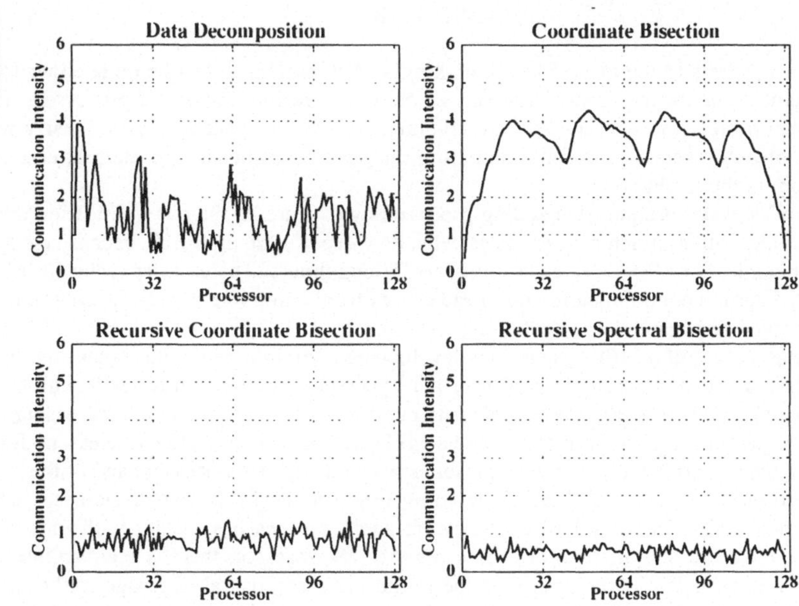

Fig. 12: Water Cooling Jacket of the 2-Valve 4-Cylinder Engine; Communication Intensity.

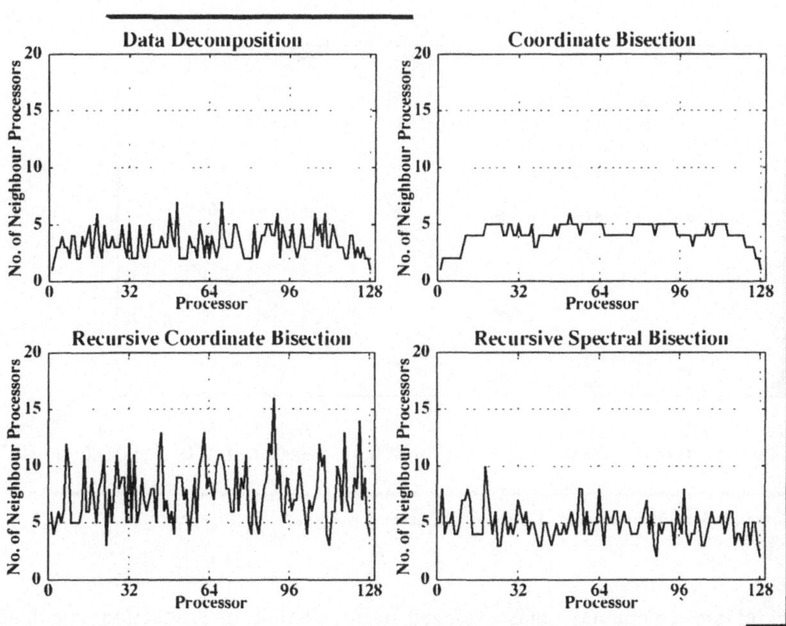

Fig 13: Water Cooling Jacket of the 2-Valve 4-Cylinder Engine; Number of Neighbour processors

BENCHMARKS

There are two aspects to the task of benchmarking a MPP system. It can be done either for a fixed problemsize or for increasing problemsize on an increasing number of processors. In a pefect parallel system (no overhead due to communication) the speedup scales linearly with the number of processors, i.e. overall computing time decreases linearly in the first case and stays constant in the second one.
In the case of constant problemsize adding processors will increase the communication intensity. In the limit, with as much processors as cells, its value will go up till 18, because the hexahedral cell elements of FIRE are each linked to 18 neighbours. Therefore speedup for a real system with a certain communication overhead ceases from being linear, because communication may become dominant over calculation.
The promising aspect of a MPP system in a developement environment with tough timescales and increasing demands for computing power is its possibility to match this needs with scalable increase of performance. Fig. 14 shows turnaround times for a problem with increasing size on a vector supercomputer, a workstation and a MPP system with up to 128 Processors. It can be seen, that the performance of the workstation is reached with 16 processors and of the YMP with 128 processors on the nCUBE2. Computing time for the MPP system increases when communication comes in from 1 to 2 processors and stays nearly constant with increasing problemsize and increasing number of processors. At 128 processors there is again an increase in computing time visible, which may be due to the increased overall message traffic in the system. Further investigations have to be done concerning this point

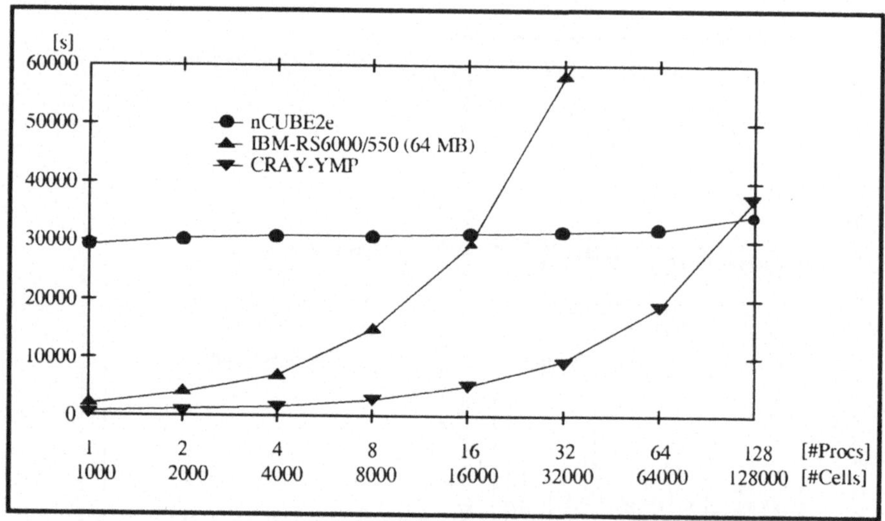

Fig. 14: Benchmark for Increasing Problemsize; Turnaround for Steady State Calculation

In Fig. 15 relative computing times averaged over a sample of production calculations at BMW are given. The size of the problems has been such that the decompositions yielded an average of about 1200 cells on each of the 128 processors of the MPP sytem. The calculations have been performed for steady state solutions where only a small fraction of the overall computing time is spent doing I/O. The speedup for the CRAY multiprocessor runs is only possible for a dedicated system. Under normal batch load autotasking enhances overall throughput rather than single job performance. There are already some improvments visible for the nCUBE2e system compared to fig. 14. Times lie slightly below the time for the YMP processor. The nCUBE2s sytem with 128 processors, installed at BMW in July, already outperforms the YMP.

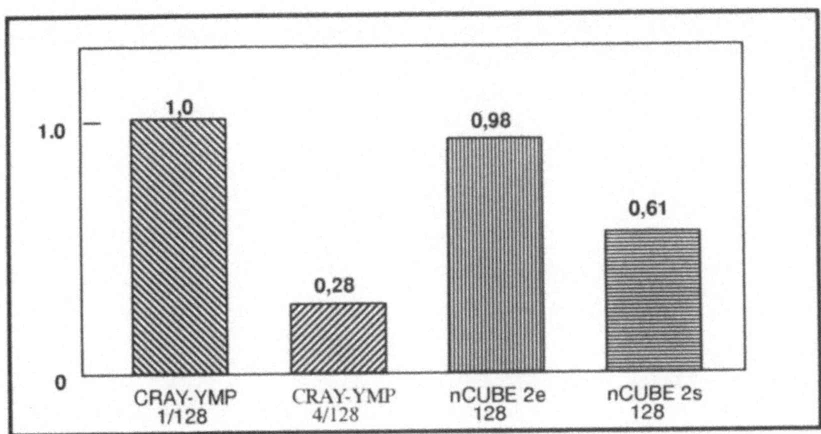

Fig. 15: Averaged Benchmark Data for Several Production Problems

Calculations for unsteady problems, producing large amounts of data, have shown that there have to be done some optimisations for the parallel I/O.

CONCLUSIONS

The MPP system at BMW has matured to a stage where it is possible to use it as a dedicated system in a productive environment in nearly the same way than conventional CPU servers. It is completly integrated in the CAE computer environment at BMW via three ethernet connections; jobs are submitted transparently via nQS and results are stored on the file server after computation. The porting of the program using the message passing concept has been straightforward and guarantees portability across a wide range of hardware platforms.
For steady state calculations the performance of the MPP has met our expectations. Experience has to be gained which decomposition is the optimal one for a given problem. For I/O intensive applications some improvements are currently implemented.

LITERATURE

[1] FISCHER, H., KNUPE, J., TROGER, C.: „Multidimensional Computersimulation of Flow and Mixing Processes in a Catalyst" (in German), VDI-Berichte Nr. 816 (1990), pp. 379-388.

[2] TATSCHL, R., REITBAUER, R., WIESER, K., FISCHER, H., TROGER, C.: „3-D Simulation of the Gas-Exchange, Mixture Preparation and Combustion Process in a State-of-the-Art 4-Valve Engine" (in German). Proceedings „The Working Process of The Internal Combustion Engine", Graz, 27./28. Sept. 1993, Editor: Pischinger R., Technical University, Graz.

[3] FISCHER, H., KNUPE, J., TROGER, C.: „FIRE on BMW's nCUBE2", SPEEDUP, Vol. 6, No. 2 (1992)

[4] SIMON, H., D. :"Partitioning of unstructured problems for parallel processing", Computing Systems in Engineering, Vol. 2 (1991), No. 2/3, pp. 135-148.

[5] FISCHER, H., TROGER.C, WIESEN, B.: „Influence of decomposition techniques and parallel I/O on the computational efficiency of MPP systems", (to be published), 1994.

DISTRIBUTED NUMERICAL SIMULATION ON WORKSTATION NETWORKS

W. Huber, R. Hüttl, M. Schneider, C. Zenger

Institut für Informatik
Lehrstuhl für Ingenieuranwendungen in der Informatik
Technische Universität München
Arcisstraße 21, D-80333 München, Germany
e-mail: huberw/huettlr/schneidm/zenger@informatik.tu-muenchen.de

SUMMARY

Networks of workstations are an interesting and valuable tool for studying and executing parallel programs. This paper investigates their use for the numerical simulation of complicated processes in various fields of technical applications. It is demonstrated that the limited communication rate in comparison to dedicated parallel computers is not always a serious bottleneck in this area of applications. Moreover, there are also advantages of this approach which in practice may be much more important than its deficiencies.

INTRODUCTION

Today, the use of a network of workstations is considered as an interesting alternative to dedicated parallel computers to enter the world of parallel computing. Several reasons support this approach:

- Networks of workstations are already in use in universities and in industry. Thus, no investment in hardware is needed to enter the business of distributed and parallel computing.
- Workstations are used only to a rather small percentage during the regular working hours and much less over night and at weekends. The use of idle times by distributed computing would improve the effectiveness of the investment.
- The users of workstations already know how to work with their workstations. They are familiar with the software tools and they know tricks to overcome the limitations and bugs of the software. Furthermore, because of the big workstation market, software and hardware are of a rather good quality and reliability in comparison to specialized parallel computers.
- Workstations are usually at the forefront of hardware and software technology. They are available with the most modern generation of processor chips, memory is rather cheap and therefore of reasonable size, and optimized compilers generate efficient code. This means that the relation of prize to efficiency is usually better for workstations than for specialized parallel computers.

On the other hand, there are also limiting factors which make this approach difficult or even impossible for many applications.

- For workstation networks, the speed and the bandwidth of communication are usually significantly inferior to the corresponding values of parallel computers. This implies that much more effort is needed in the construction of algorithms to reduce the amount of communication as far as possible. In many applications, this is very difficult and sometimes even impossible. It is one of the aims of this paper to show that there are, however, many interesting problems where communication is no serious bottleneck.
- Usually, we have incomplete knowledge about the load of workstations, because they are used by different people executing jobs of varying size like inspection of e-mail, word processing, or number crunching. This fact makes it difficult to control the parallel computation, especially with respect to a reasonable load balancing strategy. This problem is neglected in this paper. All tests were performed at night when all workstations were idle.
- It may happen that in a big network of workstations a varying number of workstations may not be accessible or may fail to do their job in the case of a distributed application. We shall not discuss possible strategies to overcome this problem.
- The software to manage a system of communicating processes in a network of workstations is not of the same quality as for other tools. Thus, e.g., the communication has to be programmed by explicit send and receive subroutine calls also in cases where such calls could be generated from some more abstract description of the computational graph (see [14], e.g.). On the other hand, available software is not as efficient as it could be. Therefore, we had to implement an efficient low level broadcast-mechanism, e.g., where the communication time is not dependent on the number of workstations to which a message is broadcasted.

In this paper, we consider three case studies for the use of a network of workstations in different areas of applications. We start with a short introduction to these applications characterizing the special features in our context. A more detailed description of the background of these problems and a discussion of test results is given in the following sections.

The first application is the solution of a problem in the area of elliptic partial differential equations. Many typical number crunching problems are of this type. The approach uses a new technique which allows not too massive parallelization, as it can be realized on networks of workstations. Furthermore, the necessary amount of communication is very small. The idea is similar to extrapolation techniques widely used in numerical analysis. A set of solutions, computed on rather coarse grids, is combined to a solution of much higher accuracy. The idea is applicable to many applications. For the most simple example of the Laplacian on a cube, it is shown in this paper that on a network of 110 HP workstations an overall rate of more than one Gflops can be achieved. Applying standard methods to the same problem, a comparable performance can be reached only on rather big and expensive supercomputers. The method was also implemented in cooperation with the GMD on a network of IBM workstations. This implementation received a third prize in the SuParCup 1993.

The second case study considers the application of standard finite element codes to typical engineering problems. The parallel implementation uses substructuring techniques where the tree of substructures is mapped to the network of workstations. A package of programs handles the distribution of stiffness matrices and components of the computed solutions to the corresponding processes distributed on the network. As a typical example, a plane stress problem is solved using adaptive refinement in those regions where some components of the solution are singular. If a reasonable distribution on the network is formed in advance before the beginning of the computation, a surprisingly good efficiency is possible reducing the computation time by a factor of 1/8 if we use 16 workstations instead of one. An advantage of this method, which may be even more important than the reduction of computer time, is the reduction of memory space: Problems which cannot be solved on one workstation because of insufficient memory may be well computable on a network of computers, because here the total memory is the sum of all local memories of the workstations which can be reserved for the application, and this may be large enough even for complicated problems. We note that in this case not only the solution process is parallelized but also the initial computation of the stiffness matrix and the postprocessing analysis. This is very natural for substructuring techniques and, in addition, supports the teamwork of a group of engineers where every engineer is responsible for a certain part of a big and complicated structure.

In the third case study we investigate a typical problem from electrical engineering in industry: the simulation of complicated circuits by a modern computational technique. Here, we slightly modified the program designed for supercomputers and used the same modified program on all workstations of a network. Only the part of the program which used most of the computing time was designed newly using a standard block elimination technique distributed on the network. Here, the amount of communication would increase with the number of workstations. But this problem could be reduced by broadcasting the blocks of the matrix instead of using point-to-point communication. Also for this application, we got good efficiency results, and it was possible to compute problems on a network of 16 workstations which could not be solved on the available supercomputer without modification of the program because of insufficient memory size.

Summarizing the results of the three case studies, we conclude that distributed computing on networks of workstations is an interesting and effective technique for many engineering applications and that this technique is the best and cheapest way to get familiar with the typical problems of parallel and distributed computing. We think that this observation holds true even to a larger extent in the future when a faster communication technology is cheaply accessible and when better tools for program development are available. This technique is also a good starting step for the use of specialized parallel computers. Software tools like PVM [1] which run on workstation networks and on parallel computers can be used to write software running on both architectures. Even if the total amount of work or an excessive amount of communication makes it mandatory to use a high performance parallel computer, a network of workstations may serve as a comfortable test environment.

PARALLEL SOLUTION OF THE LAPLACE EQUATION ON WORKSTATION NETWORKS USING SPARSE GRID METHODS

In this section, we study the parallel solution of the Laplace equation with the sparse grid combination method. First, we like to describe the algorithmic concept. It is based on the independent solution of many problems with reduced size and their linear combination. Parallelization results on a network of 110 workstations will be discussed.

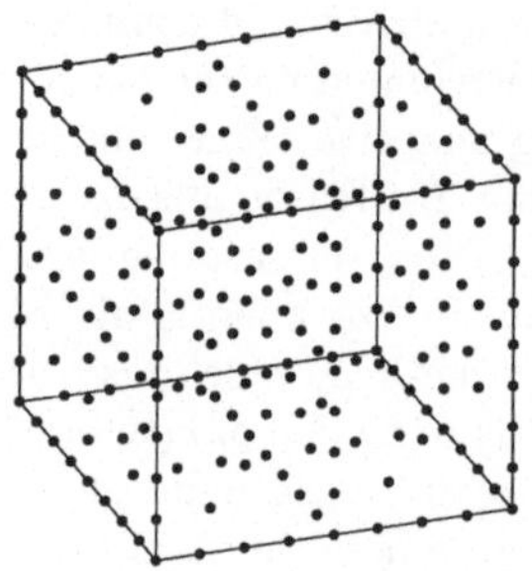

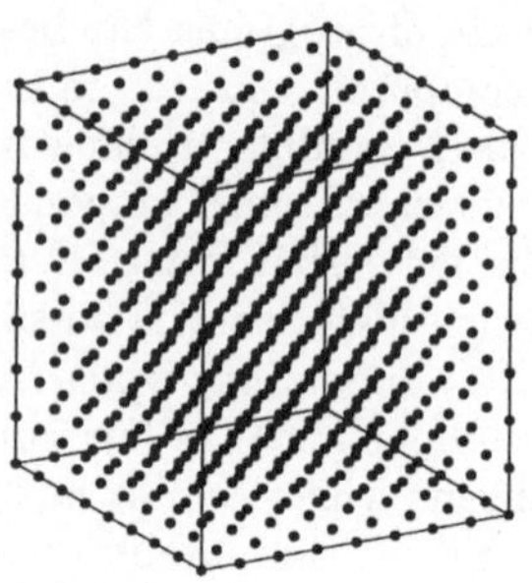

Fig. 1 The sparse grid $\Omega^s_{3,3,3}$ and the associated full grid $\Omega_{3,3,3}$.

The Combination Method

For reasons of simplicity, we consider a partial differential equation

$$Lu = f$$

on the unit cube $\Omega =]0,1[^3 \subset \mathbb{R}^3$ with a linear, elliptic operator L of second order and appropriate boundary conditions. The usual approach is to discretize the problem by a finite element, finite volume, or finite difference discretization on an equidistant grid $\Omega_{n,n,n}$ with mesh width $h_n = 2^{-n}$ in x-, y-, and z-direction, and to solve the arising linear system of equations

$$L_{n,n,n} u_{n,n,n} = f_{n,n,n}.$$

Then, we get a solution $u_{n,n,n}$ with error $e_{n,n,n} = u - u_{n,n,n}$ of the order $O(h_n^2)$, if u is sufficiently smooth. Here, we assume that $u_{n,n,n}$ represents an appropriate interpolant defined by the discrete values on grid $\Omega_{n,n,n}$. Extending this standard approach, we now study linear combinations of solutions of the problem discretized on different rectangular grids. Let $\Omega_{i,j,k}$ be a grid on Ω with mesh sizes $h_i = 2^{-i}$ in x-direction, $h_j = 2^{-j}$ in y-direction, and $h_k = 2^{-k}$ in z-direction. As introduced in [11], we define the *combined solution* $u^c_{n,n,n}$:

$$u^c_{n,n,n} := \sum_{i+j+k=n+2} u_{i,j,k} - 2 \cdot \sum_{i+j+k=n+1} u_{i,j,k} + \sum_{i+j+k=n} u_{i,j,k}. \tag{1}$$

Here, i, j, k range from 1 to n, where n is defined by the grid size. Thus, we have to solve $(n+1) \cdot n/2$ problems $L_{i,j,k} u_{i,j,k} = f_{i,j,k}$ with $i+j+k = n+2$, each with about 2^n unknowns, $n \cdot (n-1)/2$ problems $L_{i,j,k} u_{i,j,k} = f_{i,j,k}$ with $i+j+k = n+1$, each with about 2^{n-1} unknowns, and $(n-1) \cdot (n-2)/2$ problems $L_{i,j,k} u_{i,j,k} = f_{i,j,k}$ with $i+j+k = n$, each with about 2^{n-2} unknowns, and to combine their tri-linearly interpolated solutions. This leads to a solution defined on the sparse grid $\Omega^s_{n,n,n}$ (see Fig. 1). The sparse grid $\Omega^s_{n,n,n}$ is a subset of the associated full grid $\Omega_{n,n,n}$. For details concerning sparse grids, see [15] and [5].

Altogether, the combination method involves $O(h_n^{-1} \mathrm{ld}(h_n^{-1})^2)$ unknowns in contrast to $O(h_n^{-3})$ unknowns for the conventional full grid approach. Additionally, the combination solution $u^c_{n,n,n}$ is nearly as accurate as the standard solution $u_{n,n,n}$. It can be proved (see [11]) that the error satisfies

$$e^c_{n,n,n} = u - u^c_{n,n,n} = O(h_n^2 \mathrm{ld}(h_n^{-1})^2),$$

(pointwise and with respect to the L_2- and L_∞-norm), provided that the solution is sufficiently smooth. This is only slightly worse than for the associated full grid, where the error is of the order $O(h_n^2)$. Related techniques have been studied in [4] and [5].

Parallelization Aspects

Now, we study the parallelization properties of the combination method for networks of workstations. Workstation clusters can be seen as a parallel computer with high performance at low costs.

The combination method provides a straightforward way for parallelization. For our three-dimensional problem, $n \cdot (n+1)/2$ problems with about 2^n unknowns, $(n-1) \cdot n/2$ problems with about 2^{n-1} unknowns, and $(n-2) \cdot (n-1)/2$ problems with about 2^{n-2} unknowns can be solved fully in parallel (see (1)). For each of these problems, the computing time is proportional to the number of grid points on grid $\Omega_{i,j,k}$. So, a statical and very easy load balancing is possible (for details, see [9]). This parallelization potential of the combination method can be gained already on a relatively coarse grain level and makes it perfectly suited for distributed memory computers and networks of workstations.

Additionally, the subproblem solver (a multigrid solver, e.g.) can be parallelized. This, however, requires a comparatively fine grain parallelization. In addition to the natural coarse grain parallelism, this can only be exploited on massively parallel systems with thousands of processors. Alternatively, the subproblems are ideal for vectorization. In this paper, we focus on the coarse grain parallelization approach only.

Parallelization Results on a Network of Workstations

Now, we turn to the results of our numerical experiments. To simplify the presentation, we consider the simple three-dimensional Laplace model problem

$$\Delta u = 0 \quad \text{in } \Omega =]0,1[^3 \tag{2}$$

with Dirichlet boundary conditions on $\partial\bar{\Omega}$ and the unique solution

$$u = \sin(\pi x) \cdot \sin(\pi y) \cdot \sinh(\sqrt{2}\pi z)/\sinh(\sqrt{2}\pi).$$

This problem is a very hard test problem for our method. It is characterized by a high communication and a low computing time. First, we turn to the implementation of the combination method on a network of 110 HP720 workstations. The workstations are organized in 11 clusters. Each cluster consists of one disc-server and 10 discless clients, each server with 32 MByte and each client with 16MByte main memory (see Fig. 2). Table 1 shows the execution time for various numbers of workstations and various mesh sizes $h_n = 2^{-n}$ in the different directions (for more details, see [10]).

We compared these results with a vectorized version on a CRAY Y-MP4/464 on four processing units. We implemented the combination method on the CRAY in FORTRAN using the autotasking facility cft77. The parallel treatment of the different problems is indicated explicitly by the compiler directive CFPP$ CNCALL that

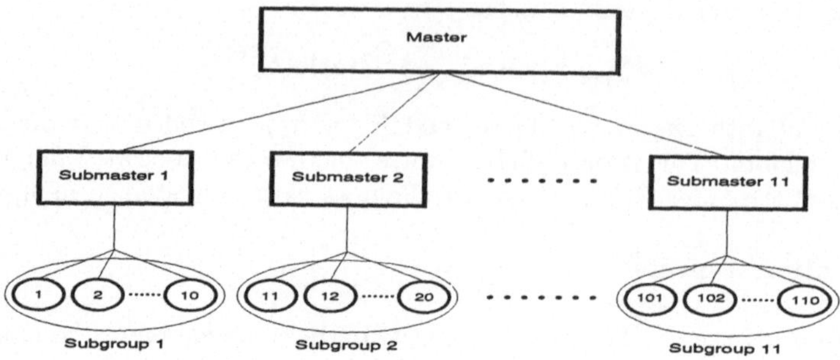

Fig. 2 Workstation network configuration.

Table 1 Times (in sec.) for the combination algorithm on $\Omega^s_{n,n,n}$ for a network with P HP720 workstations as slaves.

$P\backslash n$	4	5	6	7	8	9	10	11	12	13	14
1	0.11	0.34	1.01	2.90	8.01	21.42	55.24	139.60	351.84	943.37	2503.38
2	0.07	0.20	0.55	1.53	4.06	10.89	28.23	71.87	179.84	481.58	1265.47
4	0.05	0.12	0.29	0.83	2.07	5.50	14.33	35.54	90.09	241.57	644.21
8	-	-	0.20	0.42	1.14	2.75	7.29	18.55	45.34	123.83	328.31
16	-	-	0.12	0.22	0.61	1.45	3.64	9.14	22.93	63.69	166.25
32	-	-	0.08	0.16	0.32	0.90	2.05	4.84	12.36	36.59	89.66
64	-	-	0.24	0.31	0.44	0.67	1.38	2.68	6.63	18.16	47.45
110	-	-	-	-	-	-	1.26	2.40	4.72	12.81	30.42

exploits parallelism in *concurrentloops*. The processing units were accessible only in a multiuser environment. Therefore, the parallel efficiency is far from optimal. The run time results are shown in Table 2. Note that the 110 HP workstations are about 12 times faster than one processing unit of a CRAY Y-MP (for more details, see[10]).

Further Work

The combination method is not restricted to linear operators. Problems with a non-linear operator and PDE-systems like the Stokes equations and the Navier-Stokes equations with moderate Reynolds numbers have been solved, too (see [12]). In the future, we like to use the combination method for the simulation of turbulences. For

Table 2 Times (in sec.) for the combination algorithm on $\Omega^s_{n,n,n}$ on a CRAY Y-MP4/464 with P processing units.

$P\backslash n$	4	5	6	7	8	9	10	11	12	13	14
1	0.09	0.25	0.60	1.39	3.09	6.87	14.95	33.33	73.40	163.30	361.88
2	0.05	0.20	0.53	1.30	2.98	4.18	9.35	16.84	39.75	85.24	201.66
3	0.04	0.14	0.37	1.09	2.55	4.79	5.62	12.00	28.85	69.70	171.43
4	0.04	0.09	0.33	0.73	1.99	4.51	5.28	12.20	28.73	61.58	159.69

the extremely high amount of communication effort and main memory, more efficient communication methods and also a way to store the sparse grid solution explicitly have to be used. These aspects are discussed in [10]. In the near future, both new features will be integrated into a solver for the Navier-Stokes equations based on the combination method.

DISTRIBUTED ADAPTIVE FINITE ELEMENT COMPUTATIONS

In this section, we present an adaptive version of the Finite Element Method that is based on the principles of the recursive substructuring and, consequently, can easily be parallelized. For a test problem from the field of structural engineering, the results of a simulation on a network of loosely coupled workstations are given. A more detailed presentation of the contents of the following section can be found in [13].

The Principles of the Algorithm

The Finite Element Method (FEM) is a well-known technique, widely used for the numerical approximation of the physical behaviour of an object, see [6]. The FE algorithm that we present is based on the principle of recursive substructuring with static condensation, as it is described in [2] or [7], e.g. A computation performed thereby requires less CPU-time and main storage than a corresponding computation by the standard FEM. The recursive substructuring is a divide-and-conquer-method. The problem domain, i.e. the domain that is occupied by the analyzed body, is divided into several non-overlapping domains, the so-called subdomains, that can be recursively substructured themselves. By this, a given problem is discretized hierarchically and in a geometrically oriented way. This discretization corresponds to a substructuring tree in which every node represents a subdomain. The root of the tree corresponds to the overall problem domain and the leaves correspond to those subdomains that are not subdivided any more and that are treated as standard FEM elements in the computation. We only used quadrilateral elements with bilinear shape functions. The union of all elements forms the so-called element-mesh. During a FE computation, the data flow is alternately from the root to the leaves and back again. Therefore, a working step at any node requires the results of either its respective father-node or all its son-nodes, if existing. We used this fact in order to parallelize the algorithm.
The computations are iterated several times. After each iteration, the computed results are used to estimate the local and global error of the solution, see [3]. Based on this information, it is decided automatically which elements are to be subdivided in order to get a substructuring tree for the next run that is better adapted to the problem. A special domain description language makes the combination of recursive substructuring and adaptivity possible. The iteration stops when a predefined accuracy is reached.

The Tests

We present the results of the simulation of a problem arising in the field of structural engineering, i.e. of a quadratic panel under plane stress conditions (see Fig. 3a). The length of the panel is 1m and the thickness is 0.01m. The left side of the panel is fixed. On the upper side of the panel, there is a uniform load $p = 0.1$. The panel consists of isotropic material whose physical properties are given by the Young's modulus $E = 10^5$

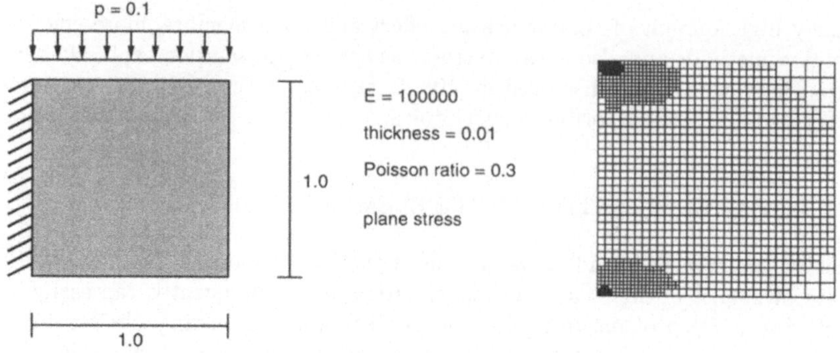

Fig. 3 Static system of the panel (a) and the FE mesh after 9 iterations (b)

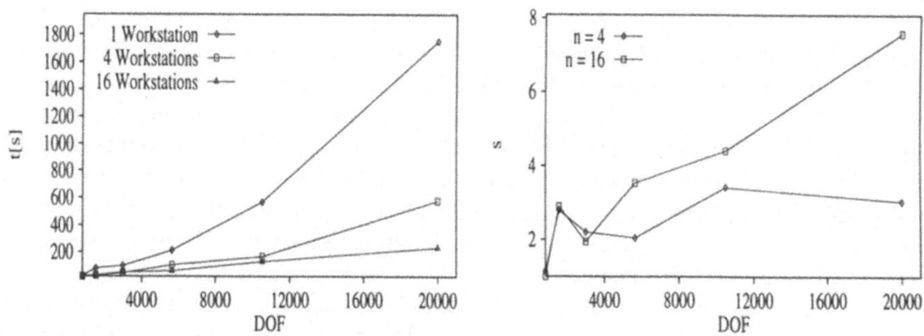

Fig. 4 Timetable (a) and Speed-up (b) for parallel adaptive FE computation

and the Poisson ratio $\nu = 0.3$.

Starting with an initial mesh of four elements and 18 degrees of freedom (DOF), an adaptive FE analysis of 11 iteration loops is done, coming up in a mesh with about 10.000 elements and 20.000 DOF. All meshes apart from the start mesh have a structure similar to that shown in Fig. 3b.

The simulations were performed on a single HP720 workstation, on 4 HP720 workstations, and on 16 HP720 workstations connected by an Ethernet with a maximum transfer rate of 10 MBit/sec. The user times necessary for the calculation with different numbers of DOF are given in Fig. 4a. The corresponding speed-up diagram is shown in Fig. 4b.

Interpretation of the Results

When four processors were used, they were distributed to the subdomains as shown in Fig. 5. Each processor performed the working steps of the nodes that were part of

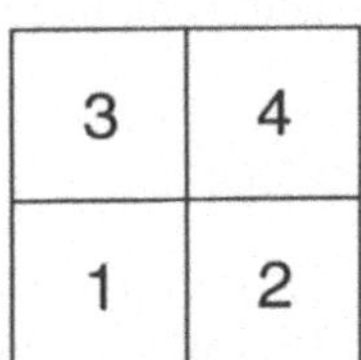

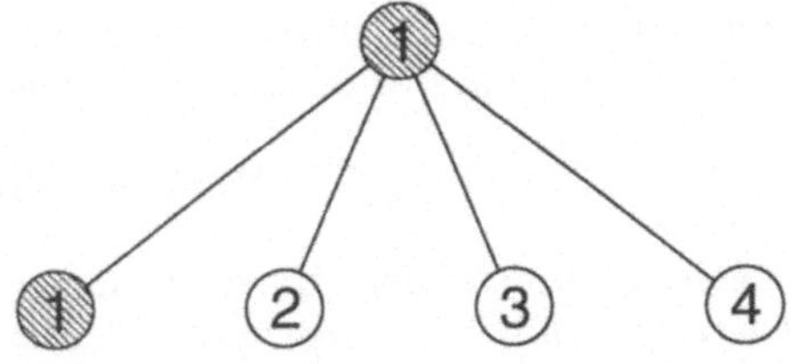

Fig. 5 Distribution of four processors to the subdomains (a) and the corresponding tree of processors (b)

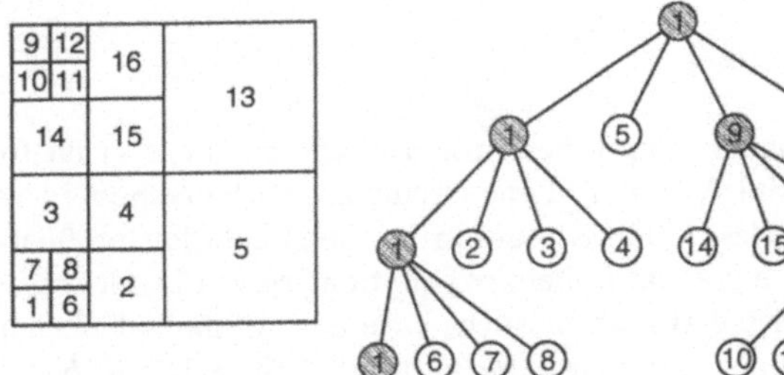

Fig. 6 Improved distribution of 16 processors and the corresponding processor tree

a subtree representing a quarter of the overall problem domain. Processor 1, in addition, computed the working steps of the root node. Thus, the processor tree in Fig. 5b corresponded to the two upper levels of the substructuring tree mentioned in the section *The Principles of the Algorithm.* Because of the adaptive structure of the mesh, the substructuring tree was strongly unbalanced. Consequently, with our contribution scheme, the processors 1 and 3 had far more to compute than the processors 2 and 4. Therefore, the speed-up that we achieved with large numbers of DOF was only about 3 and not 4 (see Fig. 4b).

Our first approach for the employing of 16 processors was to divide the problem domain into 16 equal subdomains and to assign each subdomain to one processor. We had to realize that, because of the excessive busy waiting of the processors with lower load and the arising higher communication costs, the performance on 16 workstations (not included in Fig. 4) was nearly the same as on four processors. Therefore, in a second approach, we used a processor tree that is better suited for the given problem (see Fig. 6), and by which a much better speed-up of about 8 is reached.

The different results of the two approaches show that the performance of the computations on many workstations depends essentially on the way how the processors are assigned to the substructuring tree. The building even of a relative simple contribution scheme as, e.g., the processor tree in our second approach requires a lot of time and is very error-prone, if it is done by hand. Furthermore, the structure of the arising element meshes often is not known at the beginning of the computation and may change from iteration to iteration. Therefore, in a future release, we want to use an automatically performed dynamic load balancing. Then, at the end of each iteration, according to the new substructuring tree, a contribution of the processors to the nodes can be calculated where the load on all processors is roughly the same.

DISTRIBUTED ANALOG CIRCUIT SIMULATION

The simulation of an analog circuit by the circuit simulator TITAN at the Siemens AG repeatedly creates very large sparse blockmatrix systems of equations. Until now, this kind of problem was computed either on a single workstation or on a vector computer. In this section, we present additional software to TITAN that makes it possible to simulate analog circuits on a cluster of workstations. The original program TITAN had to be changed only slightly. The content of this section is described in detail in [14].

The System of Equations

For the numerical simulation, a description of the circuit is given to TITAN. The voltages that arise at the N nodes of the circuit are to be computed as a superposition of $2K + 1$ frequencies. TITAN transforms the simulation problem by nonlinear frequency analysis to a nonlinear system of equations solved by Newton's method (see [8] or [14], e.g.). All arising system matrices have a very sparse blockmatrix-structure $(M_{ij}), \quad i, j = 1, 2, \ldots, N$, where each M_{ij} is a full $(2K + 1) \times (2K + 1)$ submatrix. However, some rows of the blockmatrices are rather full. For realistic applications, the LU-decompositions of the system matrices require more than 96% of the overall computation time.

General Parallelization of the Simulation

Let P be the number of available processors. After the distribution of the start data, on each workstation p, $p \in \{1, \ldots, P\}$, a simulation process SP_p is started in parallel. The following three initialization steps are executed by each SP_p without any net communication.

1. N is selected from the circuit description, and $I_R := \{1, \ldots, N\}$. The table $T^p \in \{-1, 0, 1\}^{N \times N}$ is created as follows: All elements M_{ij} that are non-zero at any state of the computations are indicated by $T^p_{ij} = 1$, the others by $T^p_{ij} = 0$. The value "-1" will be needed in the subsection. Because of identical input data, we have $T^p = T^q \; \forall \; p, q \in \{1, \ldots P\}$.

2. Using N and T^p, a disjunct distribution D^p of the rows of the blockmatrix to the different workstations is evaluated, see section below. Again, $D^p = D^q \; \forall \; p, q \in \{1, \ldots P\}$.

3. Each SP_p builds $I^p_R := \{i \in I_R, \; \textit{row } i \textit{ is contributed to process } p\}$.

Now, the creation and solving of the linear system is executed in four steps. These steps are iterated until convergence is achieved.

4. Each SP_p generates only the submatrices M_{ij} with $i \in I^p_R$ and the corresponding parts of the right-hand-side vector.

5. In close cooperation, the SP perform the LU-decomposition of the system matrix and, analogously, the forward/backward substitution, as explained in the next subsection.

6. Now, each SP_p has those parts of the new solution vector that correspond to a row $i \in I_R^p$. SP_1 collects the datablocks, concatenates them, and sends the resulting solution vector to all other processes.

7. All SP_s check the new solution. They come to the same decision whether to repeat the loop or not.

Parallelization of the LU-Decomposition

Normally, an LU-decomposition is done in sequential order. Algorithm 1 shows a simple way to parallelize the LU-decomposition.

Algorithm 1:

```
parallel_forall ( p ∈ {1,...,P}) :
    ⌈ for ( i = 1 to N)
        ⌈ if ( i ∈ I_R^p)
            ⌈ QR − decompose M_ii
            ⌊ send row i to all other processes
          else
            await row i                                          (1)
          forall ( j ∈ {k ∈ I_R^p, k > i})                       (2)
            if ( M_ji ≠ 0 )                                      (3)
              ⌈ M_ji := M_ji · M_ii^-1                           (4)
                forall ( k ∈ {i + 1,...,N}) :
              ⌊   if (M_ik ≠ 0)      M_jk := M_jk − M_ji · M_ik
    ⌊   ⌊
```

For the computation of M_{ji} at (4), not the inverse of M_{ii}, but its QR-decomposition is used. Talking about row i being subtracted from a row j refers to (4) and the following for-loop.

In algorithm 1, all processes are synchronized at (1), and then the j-loop is parallelized at (2). In a sparse matrix, after receiving row i, a process may very often stay idle until receiving row $i+1$ due to (3).

In the following, improved algorithm 2, the entries T_{ij}^p with $i \in I_R^p$ and $j < i$ indicate that row j needs not to be ($T_{ij}^p = 0$), still must be ($T_{ij}^p = -1$), or already has been ($T_{ij}^p = 1$) subtracted from row i.

Algorithm 2:

```
parallel_forall p ∈ {1,...,P} :
    ⌈ Initialize_for_LU-decomposition
    ⌊ Compute_LU-decomposition
```

I_d^p contains the indices of the completely computed rows of SP_p.

Initialize_For_LU-decomposition:

```
forall (i ∈ I_R^p)
    ⌈ forall (j ∈ {1,...,i − 1})
```

$if\ (T^p_{ij} = 1)\ \ T^p_{ij}\ := -1$
$\lfloor\ I^p_d := \emptyset$

In *Compute_LU-decomposition* SP_p stops when all its rows are entirely treated.
Compute_LU-decomposition :
(*Label* :) *while* $(I^p_R \setminus I^p_d \neq \emptyset)$
$\lceil$ *Phase_Of_Sending*
$\lfloor$ *Phase_Of_Subtracting*

During *Phase_Of_Sending*, the indices of those rows of SP_p from which no row must be subtracted any more, but which still do not belong to I^p_d, are united in I^p_h. These rows are determined by analyzing the entries of T^p (for details see [14]).
The diagonal blocks of these rows are QR-decomposed. For all $i \in I^p_h$, all blocks M_{ij} with $j \geq i$ are sent to all processes that need row i. These processes also are evaluated from T^p. Finally, all rows of I^p_h are included into I^p_d.

In *Phase_Of_Subtracting*, the set I^p_v is built containing the indices of all entirely computed rows which process p can dispose. I^p_v is the union of I^p_d and I^p_s, where, now, all rows get registered that have been sent to SP_p from the other processes.
Then, for an i of I^p_v, the set I^p_{su} of indices of rows from which row i now can be subtracted by SP_p is built by evaluating T^p. If I^p_{su} is not empty, i is subtracted from these rows, the corresponding entries of T^p are set to "1", and *Phase_Of_Subtracting* is quitted. This procedure guarantees that a row from which no row needs to be subtracted any more is sent immediately to those processes that wait for this row. Otherwise, i.e. if I^p_{su} is empty, i is deleted from I^p_v and, if $I^p_v \neq \emptyset$, another element from I^p_v is selected.

Often, due to the sparseness of the system matrix, in Algorithm 2 it is not necessary to perform the steps of the LU-decomposition in the sequential order shown in Algorithm 1, because at the same time from a given row i several rows can be subtracted, and the order in which these subtractions are executed is of no importance. Thus, if just one of these rows is available for the process SP_p to which row i belongs, SP_p can continue to work. Algorithm 2 detects the possibilities that arise from such concurrencies dynamically, when it builds an I^p_{su}. In this way, the idle times are reduced significantly.

Hardware/Software

The hardware we used for our tests is a cluster of 16 Hewlett-Packard 720 workstations that are connected by an Ethernet with a transfer rate of 10 MBit/sec. The distributed simulation is realized by a slightly changed version of TITAN and two short additional routines.

The Distribution of the Rows

In order to make the simulation of very large circuits possible, our implementation ensures that, on a homogeneous cluster of workstations, each computer holds the same amount of data. Other strategies may try to achieve an equal distribution of the CPU- or the hard disk load or to reduce the network load and interprocess data dependences.

All strategies can be performed after the building of T^p in step 1 of the initialization is finished. Thus, dynamic load balancing is not necessary.

Network Communication

Almost all the network communication takes place during the LU-decomposition, for which the net is a possible bottleneck. After the computations in a matrix row are finished, the diagonal block of this row and all blocks right from the diagonal block, i.e. about one half of the row, are, in typical simulations, sent to two or three processes. If direct point to point connection is applied, the number of bytes that are transmitted roughly equals

$$0.5 \times 2.5 \times (\textit{number of full entries of the blockmatrix}) \times (2K+1)^2 \times 8,$$

if each floating point number is represented in 8-Byte-accuracy. Note that the main storage need of a system matrix is given by

$$(\textit{number of full entries of the blockmatrix}) \times (2K+1)^2 \times 8.$$

If, e.g., the system matrix in total requires 200 MByte of main storage, 250 MByte have to be transmitted. Because of arising collisions and the fact that we used the rather slow UNIX rcp-command, on an Ethernet the transmission lasts between 500 and 1000 sec.. Even if the distributed LU-decomposition needs more than 1000 sec., the fact that the sending of rows is considerably delayed increases the idle times of the processes and, consequently, the total computation time significantly.
In our tests, the floating point numbers were sent in 4-Byte-accuracy in order to halve the netload on the Ethernet. This decision is based on the assumption that the corresponding rounding errors caused thereby can be balanced out by repeating the loop that consists of the steps 4, 5, 6, and 7 a few extra times. The implementation of a multicast procedure based on the UDP-protocol was started allowing that each data package has to be transferred only once, and, in the course of this one circulation, is read by all interested processes. First experiences with an experimental version let us expect that transmission with 8-Byte-accuracy will be possible with comparable or even better communication times when this procedure will be available.

Test Results

Our tests were done with a typical analog converter circuit, that consists of 86 bipolar transistors and has 247 nodes. On average, on a row only about 6 blocks are full. Row 241, however, has 84 full entries.
At first, we consider the steps of the subsection *General Parallelization of the Simulation* with the exception of the LU-decomposition. The initialization is performed in a few seconds. In step 4, i.e. the row-wise building of the system matrix, we even achieve a superlinear speed-up. The forward/backward substitution still is executed analogously to algorithm 1 and not algorithm 2 of the subsection *Parallelization of the LU-Decomposition.* This will be changed, but now, there is hardly any speed-up. The computation of a Newton iteration (including the LU-decomposition) for our test problem with $K = 59$ on 1, 2, 10, and 15 computers took 102, 62, 28, and 21 minutes, respectively. The speed-up diagram is given in figure 7.

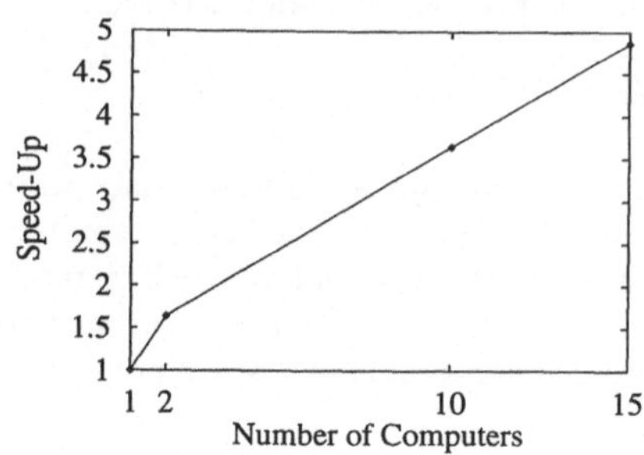

Fig. 7 Speed-up for a whole Newton-iteration

Now, we concentrate on the LU-decomposition. On a single workstation, the LU-decomposition is performed in 3400 sec.. About 1800 sec. are needed for the 3103 numerical block matrix operations (QR-decompositions, computations of $M_{ji} \cdot M_{ii}^{-1}$, matrix multiplications), 1600 sec. for block swap operations.
The simulation on two workstations required 1900 sec.. Thus, the speed-up is only 1.79. The reason for this is that one of the processes had to wait about 170 sec. for row results from its respective partner process. These idle times arise because the CPU-load is slightly unbalanced and because rows are sent over the net by the slow and hard disk intensive UNIX-rcp command.
On 15 workstations, the LU-decomposition was performed in 550 sec.. A speed-up of 6.18 and an efficiency of 41.2% was achieved. Because data are sent over the net only in 4-Byte-accuracy, the Ethernet is not the bottleneck. Instead, the bottleneck is the process to which row 241 is dedicated. At first, it computes other rows than row 241 for 60 sec.. Next, it waits 50 sec. before the rows arrive that can be subtracted from row 241. Then, the process needs 400 sec. to compute row 241. This is 8.5 times faster than the time in which a single process performs the total decomposition. More cannot be achieved, because one ninth of all matrix block operations are performed on row 241, and, since for this 350 external blocks must be swapped in, the hard disk load is about one eighth of that of a single process performing the LU-decomposition. After another 40 sec., the last row is entirely computed. Thus, since the first 110 sec. hardly can be avoided, our distribution of the rows produces the best possible speed-up.

For the LU-decomposition, the computation times on 15 workstations for different K are shown in Table 3, that also includes the corresponding memory- and CPU-needs. We emphasize that on many vector computers the solving of a linear system needing 265 or even 352 MByte of main storage is not possible.

A future implementation will not only use a main-storage-to-main-storage communication and a broadcast protocol, but as well break up the largest rows. Then, a very fast computation of problems that need more than 1 GByte of memory should become possible on a heterogeneous cluster of 100 or more workstations.

Table 3 An LU-Decomposition with different K on 15 workstations

number of frequencies K	19	29	39	49	59	69	79
computation time (sec.)	84	180	264	396	550	740	960
storage needed (MByte)	22	49	88	113	198	265	352
floating point operations (Gflop)	0.35	1.25	2.31	6.01	10.5	16.80	25.21

REFERENCES

[1] A. G. A. BEGUELIN, J. DONGARRA, R. MANCHEK, AND V. SUNDERAM, *A user's guid to PVM*, technical report, ORNL/TM-11826, Mathematical Science Section, Oak Ridge National Laboratory, 1991.

[2] L. M. ADAMS AND R. G. VOIGT, *A methodology for exploiting parallelism in the finite element process*, in NATO ASI, Vol. F7, Springer Verlag, 1984.

[3] I. BABUSKA AND W. C. RHEINBOLDT, *Error estimates for adaptive finite element computations*, in SIAM J. Num. Anal., Vol. 15, pp. 736-754, 1978.

[4] H. BUNGARTZ, *An adaptive Poisson solver using hierarchical bases and sparse grids*, in Proceedings of the IMACS International Symposium on Iterative Methods in Linear Algebra, P. de Groen and R. Beauwens, ed., Amsterdam, Elsevier, 1992.

[5] ——, *Dünne Gitter und deren Anwendung bei der adaptiven Lösung der dreidimensionalen Poisson-Gleichung*, Dissertation, Institut für Informatik, TU München, 1992.

[6] D. S. BURNETT, *Finite element analysis*, Addison-Wesley Publishing Company, 1987.

[7] R. H. DODDS AND L. A. LOPEZ, *Substructuring in linear and nonlinear analysis*, in International Journal for Numerical Methods in Engineering, Vol. 15, pp. 583-597, 1980.

[8] U. FELDMANN, U. WEVER, Q. ZHENG, R. SCHULTZ, AND H. WRIEDT, *Algorithms for modern circuit simulation*, in Archiv für Elektronik und Übertragungstechnik,Vol. 46, No. 4, pp. 274-285, 1992.

[9] M. GRIEBEL, W. HUBER, U. RÜDE, AND T. STÖRTKUHL, *The Combination Technique for Parallel Sparse-Grid-Preconditioning and -Solution of PDEs on Multiprocessor Machines and Workstation Networks*, in Proceedings of the Second Joint International Conference on Vector and Parallel Processing CONPAR/VAPP V 92, L. Bouge, M. Cosnard, Y. Robert and D. Trystram, ed., Springer Verlag, 1992. Also available as SFB Bericht 342/11/92 A.

[10] M. Griebel, W. Huber, T. Störtkuhl, and C. Zenger, *On the parallel solution of 3D PDEs on a network of workstations and on vector computers*, in Lecture Notes in Computer Science, Computer Architecture: Theory, Hardware, Software, Applications, A. Bode and M. DalCin, ed., Springer Verlag, 1993.

[11] M. Griebel, M. Schneider, and C. Zenger, *A combination technique for the solution of sparse grid problems*, in Iterative Methods in Linear Algebra, P. de Groen and R. Beauwens, ed., IMACS, Elsevier, North Holland, 1992, pp. 263–281. Also published in: SFB Bericht 342/19/90.

[12] M. Griebel and V. Thurner, *The efficient solution of fluid dynamics problems by the combination technique*, SFB Bericht 342/1/93 A, Institut für Informatik, TU München, 1993. To be published in Int. J. Num. Meth. for Heat and Fluid Flow.

[13] R. Hüttl and M. Schneider, *Parallel adaptive numerical simulation*, SFB Bericht 342/01/94 A, Institut für Informatik, TU München, 1994.

[14] M. Schneider, U. Wever, and Q. Zheng, *Solving large and sparse linear blockmatrix systems in anlog circuit simulation on a cluster of workstations*, in The Computer Journal, Vol. 36, No. 8, pp.685-689, 1993.

[15] C. Zenger, *Sparse grids*, in Parallel Algorithms for Partial Differential Equations, Proceedings of the Sixth GAMM-Seminar, Kiel, January 19-21, 1990, W. Hackbusch, ed., Braunschweig, 1991, Vieweg Verlag.

EUROPEAN DEVELOPMENTS IN HIGH PERFORMANCE COMPUTING — A COMPARISON WITH DEVELOPMENTS IN OTHER OECD COUNTRIES

D. Kimbel
OECD, 75 775 Paris; Tel: +33 1 4524 93 95
Fax: +33 1 4524 93 92; e-mail: dieter.Kimbel@OECD.Fr

SUMMARY

This paper describes the emerging concept of High Performance Computing and Communications (HPCC) and discusses its technological, scientific and economic dimensions. It then briefly reviews development trends in the United States and Japan with a view to map the state of the art in Europe. Emphasis is placed on public, national and regional promotion programmes to develop and/or implement HPCC systems for scientific applications.

The ideas presented in the paper are those of the author and do not necessarily reflect those of the organisation he is affiliated with.

INTRODUCTION: OECD AND ITS AIMS

The Organisation for Economic Co-operation and Development (OECD) is the main forum for monitoring economic trends in its 24 Member countries, the free market democracies of North America, Western Europe and the Pacific, e.g. Australia, Japan, and New Zealand. OECD's interest in the area of high performance computing is closely related in its aims and responsibilities.

In detail, these basic aims, enshrined in its founding Convention of 1960, are:

- to achieve the highest sustainable economic growth and employment; and
- to promote economic and social welfare throughout the OECD area by co-ordinating the policies of its Member countries.

With this perspective, the Secretariat produces economic surveys of each of the Member countries, statistics, analysis and policy recommendations in a vast variety of subjects, including:

- industry;
- science and technology; and
- research and development.

The interaction with public policy in these areas occurs through its committees. The topic of this symposium on HPCC, for example, is dealt with in the Committee for Information, Computer and Communications Policy (ICCP) by delegates from the Federal Ministries of Research and Technology, Economics, Posts and Telecommunications, and Länder representatives.

With this mandate, OECD's policy interest in the area of high performance computing focuses mainly on technological, economic and industrial aspects of HPC. Particular emphasis is placed on investigating the relationship among HPCC systems in terms of new scientific tools and infrastructures, and their application in research and industrial activities. This includes the development of the different HPCC components and HPCC architecture on the supply side and the downstream knock-on effects of the deployment of these systems. This interdependence between HPC applications and HPCC systems is not simply a technical matter. In fact, there is increasing evidence that access to HPCC systems and information resources is of strategic importance for all advanced research activities and significantly determines further economic and social progress. In addition, governmental initiatives to enhance this welfare increasing potential through public policy measures – given the market deficiencies in the handling of information and related technical infrastructures – are *increasingly considered again as legitimate*[1].

The policy interests in HPCC developments and their deployment have numerous dimensions.

i) Scientific research processes involve the evaluation of huge data bases and related information sources and hence depend on high performance computing as working tools. Because of the growing complexity in all scientific and industrial research questions, HPCC increasingly determines the state of the art of research, the innovation potential and the quality of the natural science disciplines and its disciples, especially in:

- engineering science;
- computer science;
- mathematics;
- physics;
- chemistry; and
- biology.

These are generally considered as the core and motor of the innovation system of advanced economies.

ii) HPC systems used in and developed for applications in these fields often become in a few years the 'working horse' for common applications in industry, commerce and private and public services and significantly determine their levels of performance. With the shift from a sellers market to a buyers market, the broad deployment of HPCC capabilities through appropriate and economically viable networks permits the systematic search for product differentiation and development of "tailored" solutions in a wide range of activities. With this foundation for the development of new products and services, an HPCC infrastructure may trigger significant growth incentives for the creation of new industries and qualified jobs in almost all branches of the economy.

[1]See also Financial Times, 27 May 1993, *Stratetic Links between Wealth and Research, and the UK's Forsight to follow Japan, Germany and th US.*

iii) Upstream, the development and testing of new HPCC systems applications significantly spurs innovations in all the technological components – and hence the well-being of the Information Technology (IT) supply industry – contributing to the development of HPC networks, including:

- electronic components;
- computers;
- workstations and terminal equipment;
- digital transmission and switching systems;
- software and mathematical algorithms;
- expert systems;
- data bases; and
- information industries.

OBJECTIVES OF THE OECD STUDY OF HPCC

Accordingly, this OECD project investigates emerging demand for new HPCC applications and new technological requirements which may affect IT demand (hardware and software). To gain such insight, a group of experts visited a number of testbeds, pilot projects and HPCC Centres of Excellence in Japan, the United States and Europe[2] and had detailed discussions with the researchers at these sites.

In addition to questions on application, specific HPCC capacity requirements and likely technological developments, the group collected information on:

- the technological, economic and social objectives underlying the different national HPCC programmes;
- the amount of public support provided for HPCC programmes, and;
- the organisational measures taken to implement these HPCC networks.

DEFINITION OF THE HPCC CONCEPT AND EMERGING DEVELOPMENTS

The project is based on a comprehensive HPCC conception. It pursues the transition from central main-frame supercomputers towards a totally decentralised, distributed computer network architecture capable of supplying by 1995/96 computing power in the teraFLOP (one trillion floating point operations per second) range. Accordingly, an HPCC Network (HPCCN) is defined as a system comprising:

[2]Sites visited in Europe: BERKOM and TUB-Kom both in Berlin; Rechenzentrum der Universität Stuttgart (RUS); Institut für Computer Anwendungen (ICA), Stuttgart; CERN Geneva.

- high performance processors with very fast interconnections;
- new software and algorithms for exploiting their potential;
- interfaces and peripheral equipment (e.g. workstations, systems for storage and running of simulations and their visualisation);
- high speed communication networks to allow efficient interworking of the system components, remote access to these systems, interaction between HPCC and data banks, and collaboration between geographically dispersed researchers, and;
- availability of highly-trained human resources to work and develop innovative applications.

In the course of these technological developments and triggered by rapidly falling computing costs, existing application areas have changed and new ones have been realized. Traditional batch and interactive applications are increasingly replaced by client-server architectures, ultra-rapid file transfer services, Computer-Aided-Design (CAD), on-line visualisation in simulation applications, etc., all of which nurture the demand for ever higher performing and more powerful computer systems.

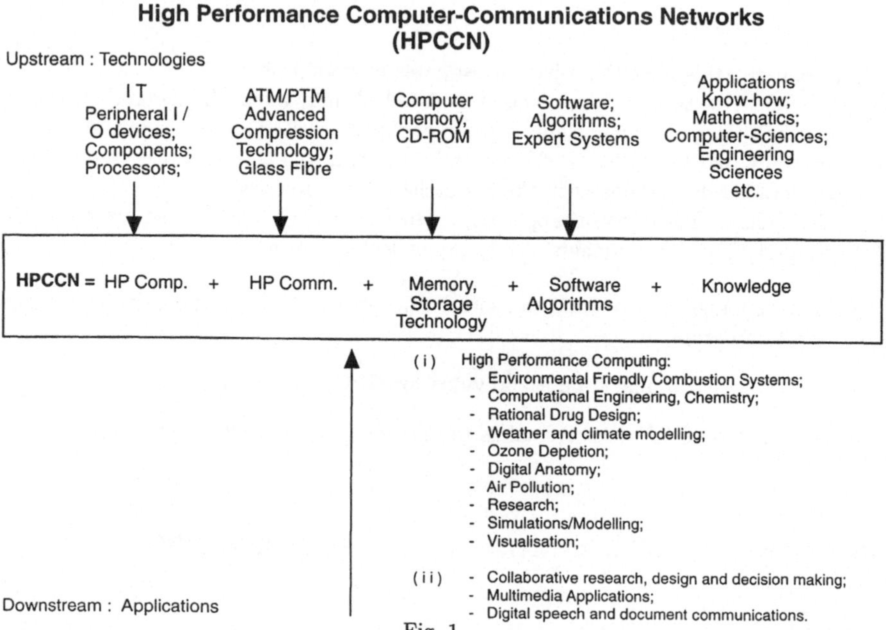

Fig. 1

The discussions at the HPCC-sites confirmed a strong correlation between applications and new information technologies (Figure 1): the requirements of emerging applications are driving the development of new IT and IT-based network configurations. The latter point is of utmost importance given the large number of components and players contributing to advanced information networks. Applications, then, determine the corridor within which IT segments develop and thus narrow the scope for pushing and pulling in different directions. Advanced applications then are also maximising the chances of getting agreement on acceptable network configurations and hence permit a larger community of researchers to interactively use these systems.

The high investment outlays for HPCC systems, however, limit their broader diffusion and use. As a consequence, small and medium-sized firms, that is, the majority of industrial establishments in OECD economies both in terms of their contribution to GNP and by the number of employees, still can not yet exploit the potential of these enabling technologies.

These high costs of comprehensive HPCC configurations, including main frame memories and external storage facilities, could of course be reduced by sharing HPCC capacities with a broader number of users. The economies of scale to result likely from distributive systems are, however, in the European countries eaten-up by the high charges for the required telecommunication services (often five times higher than comparable services in the USA) and the technical and technological barriers in high speed switching and transmission of data.

With a view to breaking this chicken and egg situation, efforts are underway in all three OECD regions focusing on the development of new communications technologies and concepts. The objective then is to bring the capacity of communication networks, i.e. the infrastructure and vehicle of HPC, with regard to quality, speed, and flexibility to the level of high performance computing and its applications.

From the site visits, it became clear that research and experiments are directed towards removing these bottlenecks in three directions:

- first, through upgrading the performance of ordinary analogue telephone circuits (ISDN strategy);
- second, by integrating different hierarchies of networks, including telephone, cable TV, and video networks; and
- third, by developing and implementing fully digital switched flexible broadband networks (bandwidth à la carte) and transmission speeds in the gigabit range (broadband-ISDN strategy).

The last strategy is the really innovative approach and an international race has started to establish an industrial research and production base in this field. The key to providing the functions required in the high speed network approach is, on the technical side, Asynchronous Transfer Mode (ATM) and, on the economic side, competition to make sure that the new system is deployed on time in the market.

ATM results from a number of technological achievements, including optical fibre cables and Large Scale Integration (LSI) technology. With the introduction of optical fibre into communication networks, transmission errors diminish. This enables the transport network to dedicate itself entirely to information transfer by delegating most flow and error control to the terminal or a computer. Advances in LSI technology enable systems to process protocols and perform switching economically and automatically without software control. In ATM networks, digitised voice, data, and video signals are divided into pre-defined 53-byte blocks (cells). The system gives each cell a header with a destination. While the cells resemble conventional packets, the fixed length simplifies the protocol so transmission can take place on a hardware, rather than a software basis[3]. Hence, computer communication convergence is finally accomplished with storage, processing, switching and transmission as integrated functions in a fully digital environment.

HPCC – AN INTERNATIONAL CHALLENGE TO TECHNLOGICAL AND ECONOMIC POLICY

The transition from the supercomputer generation of the 1970s to the HPCC concept of today mainly reflects the increasing computing requirements of scientific and industrial applications. HPC is increasingly needed to react more rapidly to changes in production and consumption patterns and to meet pressing environmental demands. The search for innovative solutions in scientific research and industry is largely made possible by HPCC supported visualisation, simulation and optimisation models. Hand in hand with these trends are a significant growth in simultaneous collaborative research between geographically dispersed researchers, and a growing demand for access to data bases and related information resources. Taken together, these developments trigger the exponentially growing demand for distributed high performance computing facilities. HPCC is not then simply a technical matter. Even the currently available prototype systems are economically viable and demonstrate well their strategic importance even in private (as opposed to military) research and development activities.

Numerous applications document that HPCCN is used in firms to decrease R&D costs and shorten the time to market of products and services. In addition, the potential of HPC to rapidly develop and test solutions for changing market and environmental conditions is increasingly determining the leading-edge in international competition. The automobile and the aviation industries therefore are already extensive users of HPCC. As an enabling tool for conceptual and lateral thinking, for the rapid testing of working hypotheses and for related scientific problems (What if...? questions), HPCC is gaining increasing momentum in almost all scientific disciplines, including mathematics, physics, biology, medicine and engineering sciences.

HPCC can also assist strategy formulation and decision-making in social and economic policy. It permits the development of alternative solutions and the visualisation of knock-on effects of the respective consequences.

In the US industrial sector alone, annual gains in productivity ranging from 1 to 3 percentage points are forecast to result from HPCCN[4]. A study conducted for the Commission of the European Communitites' (CEC) DG XIII in 1993 yielded similar results. Based on a systematic sectoral analysis of the more industrially developed regions, sophisticated information systems are predicted to increase performance by 4 to 6 per cent over the next 15 years if infrastructure implementation and usage innovations are supported.

A number of insightful scientists, however, argue that in many cases the use of supercomputers had only scratched the surface of the kinds of problems whose solution

[3]For further details see *Switching Technology for Broadband ISDN*, Chiaki Hishinuma, Executive Manager, Research Planning Depatrment, NTT Communication Switching Laboratories, in Japan Computer Quarterly, No. 91, 1992

[4]See Gartner group, Inc., *HPCC: Investment in American Competitivity*, quoted from Rubbia. See also ICCP No. 30, page 22-26.

could have a dramatic effect in the relevant field. For example, in the case of aerodynamics, though the aerodynamic characteristics of various parts of an aircraft structure could be studied in real-time using supercomputers, the study of whole aircraft designs, using the supercomputer as a computational wind tunnel, was beyond even the most powerful machine. To do this, supercomputers with speeds of thousands of times those of the fastest in existence would be needed. But the study of aircraft designs this way could lead to the elimination of both wind tunnels and also the construction of extremely expensive prototypes. It has been reported that Boeing aims at designing its next generation of aircrafts totally using supercomputers without any prototype or wind tunnel testing. Clearly the economic consequences of being able to do so would be enormous.

The challenge of doing this became known as a Grand Challenge and comparable Grand Challenges in many other fields were identified as targets for the American HPCC programme, including combustion systems, computational chemistry, ocean science, pharmaceutical design, economic modelling, improved oil and gas recovery, molecular biology, materials science, nuclear fusion reactor design and simulation, climate modelling, semiconductor design, structural engineering, superconductivity, resource and environmental modelling.

The characteristics of Grand Challenge problems are that their solution requires computers capable of speeds of at least one teraflop (= one trillion floating point operations per second), amounts of main memory in excess of one terabyte (= one million megabytes, one megabyte = one million bytes), and vast amounts of back up storage as well as an extensive high speed communications infrastructure. To compare, well-balanced supercomputers today rarely have a peak speed in excess of ten gigaflops (one gigaflop = 10^12). It is claimed that the solution to these problems will lead to a major increase in scientific or industrial competitiveness in these areas and to an enhanced ability to manage natural resources and environmental change (Figure 2).

HPCC DEVELOPMENT TRENDS AND PUBLIC PROMOTION PROGRAMMES IN JAPAN, THE UNITED STATES AND EUROPE.

The pressure to continuously search for innovative solutions in scientific and industrial activities involves the processing of ever growing volumes of data from series of experiments, daily observations, and the analysis of increasingly complex statements of problems.

Traditional central processors reach their limits here. In addition, visualisation and simulation models of fluid processes and the growing need to tackle research projects collaboratively with a multitude of geographically dispersed researchers create new requirements with respect to the architecture and access to HPC systems. The experience gained at CERN, to bring HPCC and access to data bases to the researchers (as distinct from bringing researchers to the machine) is stimulating the development of new HPCC concepts.

Meeting these emerging processing requirements at economically viable costs clearly requires breaking new technological ground in all of the HPCC components, including:

- processors and storage media;

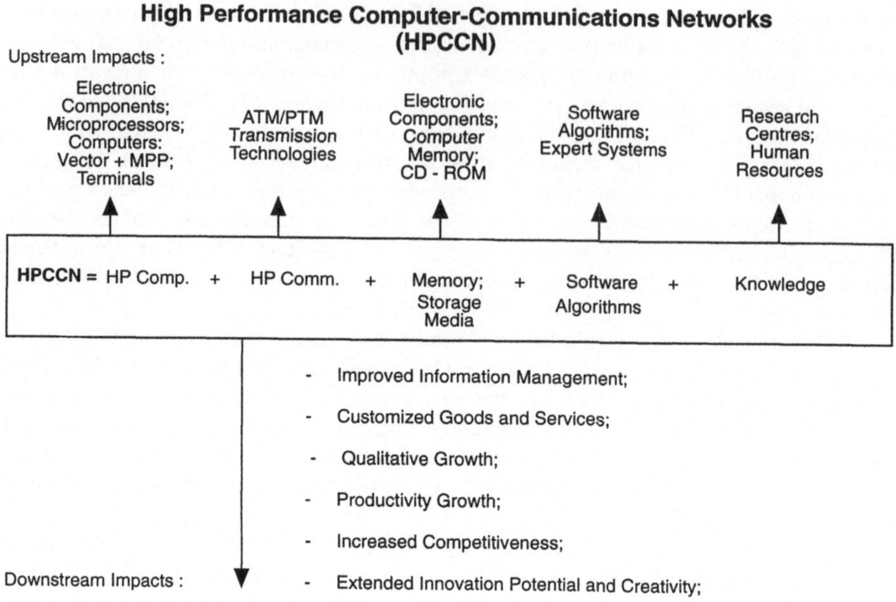

High Performance Computer-Communications Networks (HPCCN)

Upstream Impacts :

Electronic Components; Microprocessors; Computers: Vector + MPP; Terminals	ATM/PTM Transmission Technologies	Electronic Components; Computer Memory; CD - ROM	Software Algorithms; Expert Systems	Research Centres; Human Resources
↑	↑	↑	↑	↑

HPCCN = HP Comp. + HP Comm. + Memory; Storage Media + Software Algorithms + Knowledge

Downstream Impacts :

- Improved Information Management;
- Customized Goods and Services;
- Qualitative Growth;
- Productivity Growth;
- Increased Competitiveness;
- Extended Innovation Potential and Creativity;

Fig. 2

- software and algorithms;
- workstations;
- transmission and switching systems; and
- human resources.

The following discussion presents development trends and their characteristics in Japan, the United States and Europe.

Japan

R&D activities are predominatly undertaken in private firms and in particular within the different corporate families (keiretsu) and less so in the universities. Accordingly, the ratio of industry to university R&D expenditure at the national level is about 85:15. HPCC developments and in particular the application of HPCC systems follow a similar distribution pattern. Computer networks for the scientific research community, common in the US and Europe, are accordingly less deployed. In particular, what is lagging in Japan is an easily accessible, ubiquitous Internet-style TCP/IP network[5] infrastructure. Internet connectivity in Japan largely continues to be managed for the most part by individual academic institutions (e.g. the National Centre for Scientific Information Systems, or NACSIS, which has a T 1 backbone plus local trunk lines, 40 hosts, and speeds ranging from 64 to 512 kl bits) with limited government support.

[5]TCP = Transmission Control Protocol; IP = Internet Protocol.

Though ambitious plans for a Ministry of Education-sponsored IP network were discussed, little seems to have been done yet. Japan seems to be behind both the United States and Europe (NSF-Net, DFN, DWN, Janet, Renater, GARR, etc.) in this regard. However, commercial data networks are of a high quality. With the end of the telecommunications monopoly in 1985, the deployment of new technologies grew rapidly: in July 1992, some 90 000 basic rate access lines and 2800 public telephones already had ISDN plugs. In addition, the Ministry of Posts and Telecommunications (MPT) and the NTT have published an aggressive programme of investment (US$ 300 billion) in broadband ISDN (based on ATM and Glassfiber) to provide services to all businesses and virtually all homes by the year 2015.

The Government's recently announced US$ 117 billion economic recovery plan also will support the establishment of the above B-ISDN project. The demand for switched high speed networks (T1 backbone, with resale possibility for excess capacities) of private industry is already met by the Carrier Type II Organisations (e.g. a private corporate network, jointly owned by the using industry).

With regard to HPC, Japan ranks second behind the United States. The Ministry of International Trade and Industry (MITI) has dedicated plans to further strengthen the Japanese position in the world market. Its Real World Computing (RWC) programme provides some 60 billion yen (US$ 500 million) over a period of ten years to enhance vector and massive parallel processing computers.

These public R&D programmes pursue a technology-push strategy to extend Japan's research and industrial production capabilities in the HPCC area and its different components. It is however surprising that there is no integrated approach to HPCC networks. The respective R&D and investment programmes for new computer architectures and digital broadband networks, despite the fact that they both use ATM and Glassfiber, are separately implemented. MITI and MPT each pursue their respective objectives and even co-ordination among the supply industry is not sought.

In addition, no reference was made to experimenting with advanced applications in the discussions on pilot projects. The research team met all but one HPC application researcher who is supplying HPCC services as an entrepreneur (Kunio Kuwahara).

Since the Japanese HPCC concept basically pursues a hardware approach to HPCC, the lack of co-ordination among government institutions and the HPCC suppliers and users, the particularities of Japanese industrial structures and the different interests at stake may eventually delay the development of a compatible HPCC infrastructure.

The United States

Comprehensive research and production capabilities have developed in the context of the US Government's space and defense programmes for the computer and communications industry. The industry is the world leader in HP computers and related components. The evolution of high performance networking in the United States has been intimately linked to the creation of the Internet, a TCP/IP-based computer communications network that began as a research project – the ARPAnet – funded by the Department of Defence. In recent years, the standards and protocols developed for that initial experiment have been further augmented and deployed as the core of a large scale network linking US academic institutions partly financed by the National Science Foundation. Today, the Internet is a loose collection of interconnected IP networks involving over 1 000 000 computers, linked by a core backbone that continues to receive significant support from the NSF. The Internet today connects some 3 million government and academic computer users, along with an increasing number of commercial enterprises and foreign users (growing at 12% per month). One unique feature of the US Internet is the manner in which it was constructed. This packet-switched communications network was created as an entity that was functionally independent of the public switched telephone network though public communications carriers can provide private data transport services on different links within this collection of networks. Because it was conceived as effectively separate from the public switched telephone network and primarily designed to serve the government, non-profit, and educational community, it evolved free of the regulatory and other constraints faced by the public telephone network. As a testbed for developing new communications network and application technology, it also permitted a degree of experimentation that would probably not have been available to developers working on the commercial, public switched network.

Many services first developed for experimental use on the Internet are now increasingly in wide commercial use. It is probably fair to say that the current widespread use in the United States of electronic mail, file transfers among widely distributed computers and remote log-ins to physically distant computers owes much to the accelerated development of the Internet. New services and applications such as remote retrieval of high quality text and image data (digital libraries), and distributed information servers (like the Gopher[6] and WAIS applications) are only the most visible edge of a whole new generation of applications that will consume the increasing amounts of bandwidth available on the Internet as it is modernised and upgraded.

Current national efforts to upgrade communications infrastructure in the United States take two distinct paths. The HPCC Initiative (HPCCI), in addition to funding R&D on new computer architectures and software algorithms, puts significant resources into a National Research and Education Network (NREN). Other resources are being invested in a number of testbed projects created to experiment with the technologies used in very high speed (gigabit) networks. These government supported efforts focus on investment in understanding the problems and creating the basic technology required to deal with gigabit data rates, and supporting major growth in the networks linking educational and research computer users.

The second major set of initiatives involves the upgrading of the public switched telephone network. One major side effect of the deregulation of the long distance telephone market in the United States has been the entry of a significant number of new companies – including cable TV operators – into this business, and the creation of a highly competitive business environment. Not only conventional long distance voice telephony, but also digital leased lines have plunged in price. A substantial cheapening of digital data bandwidth has clearly been one of the factors driving the extraordinary growth in wide area networking in the United States. At the end of 1991, for example, US local exchange carriers had installed about 11 000 miles of optical fibres carrying data at DS-3 rates (45 megabits per second or greater), which terminated on customer premises.

Generally, circuits delivering these data rates are unavailable commercially in Japan or Europe.

---

[6]A "gopher" is originally a colloquial American term for an assistant who runs errands: "Go for (and find) ..."; and WAIS (Wide Area Information Server), applications demonstrated to the OECD expert group at Cornell University at Ithaca, N.Y.).

Currently, the major public communications carriers are offering a variety of new digital data transport services, including so-called fractional T-1, frame relay packet-switched services and Switched Multimegabit Data Services (SMDS). Many carriers have also now announced the introduction of services based on Asynchronous Transfer Mode (ATM) into their networks. ATM technology allows both voice and other data to be prioritised and transmitted over a single communications system at rates up to or eventually exceeding a gigabit per second, and promises to be the uniting technology which brings together the now separate world of Internet-style packet-switched networks with the public switched telephone network, in a single seamless, ubiquitous network.

For reintegrated research and development in enhanced high performance computing (Teraflop and beyond) and communication networks (gigabit range) to provide the tools for the Grand Challenges, the government launched the NREN Programme in 1991. Ultimately, this programme will lead to the establishment of a National Research and Communications Network to interconnect some 250 universities and major national and private research laboratories in the US. Currently, a number of testbed experiments are underway to determine the future requirements of these institutions and hence to assist in capacity planning and development of such gigabit networks and their components. Figure 3 identifies these testbeds and shows their geographic deployment in the United States.

GIGABIT TESTBEDS IN THE UNITED STATES

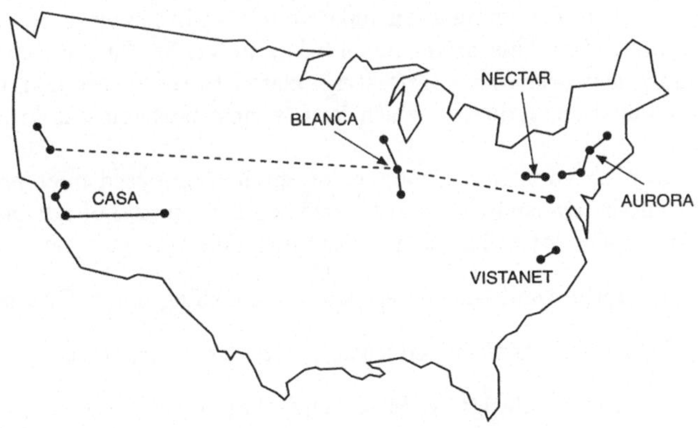

Source : Status Report on the European Gigabit Initiative, R. Popescu-Zeletin, DETEBERKOM / GMD, Berlin 1993, p.1.

Fig. 3

Europe

In general, the members of the OECD research team were impressed with the strong research base and impressive laboratory and demonstration projects observed in Germany. The experimental applications viewed reflected a high level of quality and innovation.

However, German and CERN network users in Switzerland complained about the high cost and the limited availability of high performance digital data transport facilities on the public (telephone) network in Europe. Basic leased data circuits, generally supplied by PTTs holding a monopoly position, are commonly up to five times the price of equivalent services in the US[7]. Unlike the case in the United States, a packet-switched IP network infrastructure independent of the Public Telephone and Telegraph (PTT) companies was not allowed to develop. As a consequence, although the Internet has grown rapidly in Europe in terms of nodes and connectivity, the bandwidth connecting these nodes has grown only pitifully, and high performance computer users are quick to complain about the low effective communication rates with other sites in Europe.

In February 1993, a plan for a pan-European backbone for TCP/IP and X.25 packet-switched traffic, offering speeds up to 2 megabits per second, was announced by the European Community. Although a much-needed upgrade to Europe's network infrastructure, it falls far short of the 45 megabit per second links already used on the backbone of the American NREN.

The fundamental obstacle to rapid improvement in performance of the network infrastructure in European countries would appear to be the attitudes of European PTTs who see no significant commercial market for data transport services offered at very high bit rates. Coupled with the resistance of supercomputer centres and other high performance computer users to paying very high tariffs for relatively limited bandwidth, this creates a chicken-and-egg problem of sorts.

One promising route for breaking this impasse appears to be visible in the United Kingdom. There, SUPER JANET (the British research Internet) has rapidly upgraded the cross-section of its communication links at relatively low cost, as some degree of competition now exists. This drives home a key point: in Europe, rapid progress in high performance networks seems intimately related to the future market structure of the data communications industry, which for the most part remains firmly in the hold of monopolies.

In some European countries, a number of application-based pilot projects are underway exploring the transmission and switching requirements of advanced computer-communication applications. The sites visited included:

- BERKOM, Berlin (multi-media electronic publishing, CIM, City planning);
- TUBKOM, Berlin (medicine, education, network management);
- RUS, Stuttgart (simulation, CIM, CAD, Collaborative Research, etc.);
- ICA, Stuttgart (CIM, Visualisation, Fluid Dynamics, etc.); and
- CERN, Geneva (Distributed Computing, Information Resources, etc.).

As infrastructure services still fall under national responsibility, the situation with cross-border networks, such as HEP (High Energy Physics) and EARN (European Academic Research Network), is particularly difficult. In the context of EUREKA, COSINE (Co-operation of Open Systems Interconnenction Networks in Europe) has been mandated to improve this situation. Similarly, for higher speed networks to meet the future requirements of the European scientific community, RARE (Réseaux Associées pour la

[7] See the 1993 Forum Engelberg's *Energy and Enviroment: A Question of Survival.*

Recherche Européene) is developing new concepts. However, neither is supported by real-world testbeds.

The main initiative on a Europe-wide scale is the RACE Programme (Research and Development in Advanced Communications Technologies in Europe) of the European Community. RACE aims to develop and operate high performance networks by 1995. At this stage, discussions with the different European Telecommunications Operators (TOs) foresee three networks:

- EBIT (European Broadband Interconnection Trial);
- METRAN (Managed European Transmission Network); and
- GEN (Global European Network).

The tentative network layout of the last (though the list of the members of the R&D and user community is not finalised) is presented in Figure 4.

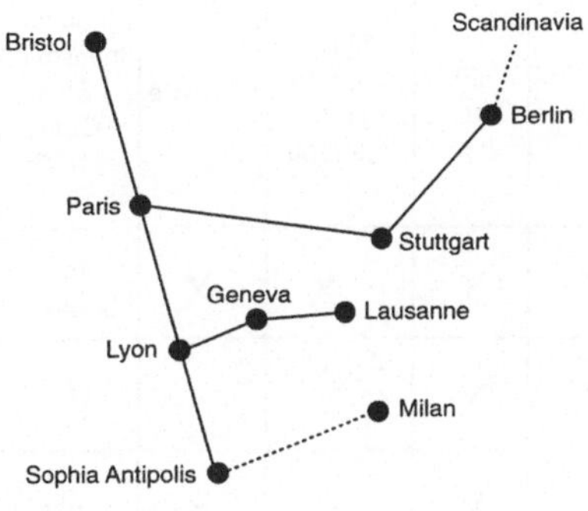

Source : Status Report on the European Gigabit Initiative, op. cit., p. 47.

Fig. 4

The European situation then can be summarized as follows:

- There are numerous important projects and field trials on partial aspects of the HPCC concept;
- Given the absence of indigenous supercomputer production capabilities, these efforts focus primarily on the development of high performance networks to permit switched, application-driven communications between workstations, computers, data bases and users.

- The respective national carriers or telecommunications operating companies are in charge of the establishment of these field trials; as these TOs, however, do not see a real demand for higher speeds, they show a rather defensive attitude toward HPCC. Consequently, even the field trials, with regard to transmission speed and bandwidth are narrowly defined: "broadband" seems to start as 34 Mbit/s compared to 600 Mbit/s in the United States.

In addition, there is no long-term, political programme in the area of HPCC as those projects underway are not focused and are isolated from related activities in the different countries. The most prominent effort towards a European programme is the Rubbia Report commissioned by the CEC. This report suggests a ten year programme for HPCC and networked systems including the full spectre of HPCC relevant components. It proposes an annual investment budget of ECU 1 billion over a period of ten years to be financed by public, private, national and European sources. If this proposal were to be endorsed, it would indeed be comparable in scope and structure to the US programme (see Figure 5).

HPCC Policy Focus and Public Funding

HPCC Components / Countries/ Regions	HPCC Applic-ations	HP Comp.	HP Comm.	Main-Memory, External Storage	Software	Industrial Investment Plans (billions of US$)	Research + HPCC - Centers of Excellence	Public Funding (billion of US-$)
USA	X	X	X	X	X	450(1) (1993-2015)	X	3 billion (1991-1995)
Japan		X	X			250(2)		0,6 (1992-2002) RWC
Europa	X		X		X	n. a.	X	Rubbia Report : 3,5 billion ECU

1. An " Infostructure" For All Americans: Creating Growth in the 21st Century, April 1993, Ameritech, Bell Atlantic, Bell South, NYNEX, Pacific Telesis, Southwestern Bell, US West, page 3.

2. NTT Investment Plans, Tokyo, July 1992.

Fig. 5

CONCLUDING OBSERVATIONS

- The scientific and industrial viability of the HPCC concept is recognized by experts from a multitude of areas (including IT/HPCC suppliers, scientific and industrial users and system providers) in the three basic OECD regions.

- The development towards switched HP computer communications networks is demand driven despite the fact that these networks are still in an embryonic stage.
- The current and emerging benefits are derived through the simultaneous interaction among research, scientific-industrial use and the production of the HPCC components and integrated HPCC systems. With appropriate policy guidance the current 'island-solutions' may develop into an HPCC-based information infrastructure. The likely macro-economic benefits of this development towards computer-communications-based information infrastructures have at least four dimensions:

  i) they make HPCC applications, including simulation of fluid processes, visualisation, simultaneous collaborative research, etc. economically affordable and a strategic tool in the tackling of scientific and industrial problems. In facilitating the development of theoretical concepts and physical experiments and in fact providing substitutes for these traditional (and related) R&D processes, HPCC is becoming indispensible for scientific, industrial, and socio-economic innovation processes;

  ii) they extend the creativy and innovation capacity of science and research including scientific education and, in addition, may yield economies of scale and synergy effects through 'networked', interdisciplanary collaborative work;

  iii) they provide new innovation incentives (resulting from advanced applications) upstream for the development of the different IT segments (electronic components, mircoprocessors, computers, digital transmission and switching systems, software, artificial intelligence, etc.) and other technologies; and

  iv) HPCC networks may become the foundation for a new telecommunications paradigm.

The enabling potential of HPCC and the broader benefits to be generated through HPCC applications are plausible and seem realisable.

As there is nothing automatic in the transformation of technological opportunities into market products and given, the market deficiencies involved in network externalities and the development of information infrastructures, there is a good argument for integrating HPCC into a comprehensive IT-based technological policy.

At this stage, however, the implementation of HPCC infrastructures basically involves conquering unexplored territories. There is no expertise in establishing ATM/PTM-based switched HPC networks and related high speed, digital general service networks. In addition, there is no reference information on the acceptance, viable pricing of services, likely usage patterns and related operational, technological and organisational aspects of HPCC.

With a view to avoiding undesirable developments and unsuccessful technological investments and gaining more insight into the basic questions, new developments and implementation strategies are required. This report suggests to interconnect HPCC pilot applications for the scientific and industrial research community to build trial networks for testing and experimenting with the different HPCC components and network architectures. This strategy of learning by experimenting at real world application sites would permit the dismantling of complex infrastructure projects into more 'visible' and scrutinisible modules and phases, and hence proceed in a step by step mode both in the development of new PH IT-based equipment and the implementation of systems and network architectures.

Candidate applications to be tested in such field trials might, in addition to Grand Challenge projects, include critical public services (postal services, transport sector, health care, education and training etc.). In fact, traditional telecommunications services themselves – given the convergence between narrow and broadband services and the increasing interaction between wired and mobile networks – might be an ideal testbed for HPCC. The objective then would be to explore whether the ATM-glassfiber network has the potential to become the "integrated solution" and whether both traditional and new services could be provided at relativly lower costs.

The results to be derived from these field trials might well contribute to facilitating both the financing and extending of such demonstration trials and send clear signals to the market to test new HPCC applications and start implementing new sections (phases) within the comprehensive concept of HPCC infrastuctures.

To trigger this process in the field of HPCC, the OECD project has developed some guidelines for the establishments of HPCC-application trial projects. Some are listed below.

- In the setting up of public HPCC application trials, the centre (often a public/private consortium) and hence their researchers, should be supplied with the best HPC system available and permitted to experiment with specific applications using alternative HPCC architectures.
- Applications specific HPCC centres might best be built around different HPCC equipment (to be readily purchased or developed for the application) involving competing suppliers.
- Trial centres should be networked among each other and provide broadband connections to related research institutions (private and/or public), individual leading researchers, and data bases in a hierarchy of levels: nationally, regionally (in terms of OECD regions: North America, Europe, and Pacific), and finally interregionally. The telecommunications operating companies, keen to learn from real world experiments, might be partners in developing the HP links, eventually to be subsidized out of public R&D budgets, without being in the lead in the management of the trials[8].

Ultimately, it might be appropriate to integrate the results generated in the different field trials to serve as a foundation for a comprehensive, medium term, application-driven technology policy strategy, at national and regional level, with an aim to intra-regional co-ordination.

[8]For the layout of the network of (application-specific and interdisciplinary applications) trial network see also Figure 3: Planned European HPC Network. Regarding the issue of which is in charge of the field trials, a number of currently applied models are worth analyzing: Internet, Consortium for National Research Initiatives (CNRI) in the US, SUPERJANET, the UK, RUS and ICA at the University of Stuttgart, FRG.

HIGH-PERFORMANCE COMPUTING IN FLUID MECHANICS

E. Krause
Aerodynamisches Institut, RWTH Aachen
Wüllnerstr. zw. 5 u. 7, 52062 Aachen, Germany

SUMMARY

The continuous progress in the development of solution techniques for time dependent, three-dimensional flow problems is unthinkable without the substantial increase of computational speed and storage capacity of high-performance computers and the remarkable development of discretization, solution, and visualization methods. This paper summarizes some of the major results obtained in the past few years at the Aerodynamisches Institut of the RWTH Aachen in numerically simulating flows in technical apparatus, biological systems, aerodynamic flows, parallel computing, and in simulating complex flows with vortical structures. Comparison to experimental data is sought whenever possible. It will be shown, that direct numerical simulation of complex flow phenomena is successful, if sufficient space and time resolution can be provided.

INTRODUCTION

With the advent of high-performance computers detailed investigation of complex flow phenomena, in particular of time dependent and three-dimensional flows, is advancing at a pace, never expected in the past, not even in the recent past. Despite this unexpected success of numerical predictions, it must be realized that solutions to the key problems in fluid mechanics, transition to turbulent flow and turbulent flow itself, are in reach only in more or less exceptional cases. This is so, since the time and also the length scales, which can be resolved in numerical predictions, are limited by the step sizes, which, in turn, are dictated by the storage capacity and computational speed. However, today we are able to capture already relatively small vortex structures with presently available solutions, and it is truly amazing, how many details of vortical flows can be predicted, provided the step sizes are sufficiently small and - even more important - if the numerical damping does not spoil the true physical picture of the flow. Since it is seldom known in numerical flow investigations, how close one is to physical reality, experimental work, more refined, and more to the point than ever before, is necessary today and in the future. We need experiments, for which in- and outflow conditions are clearly specified. We need complete experiments, that is to say, all data necessary for comparison with numerical predictions must be provided. This is still an enormous problem for time dependent flows, since handling of large amounts of data, necessary for the comparison, may be an unsurmountable problem.

For the reasons mentioned flows which are known to generate large vortex structures are preferred candidates for numerical simulations. As an example, the vortical flow around delta wings is mentioned. With the axes of the two primary vortices on the upper side of the wing aligned with the main flow direction, they seem to be ideally suited for numerical investigation. Critics often offered the view, that the breakdown process of the primary vortices, that is the sudden destruction of the vortex core, could never emerge from numerical predictions, as could not the formation of the secondary and tertiary vortices. That point of view was shown to be too conservative, and, except for transition to turbulent flow and the turbulent part of the boundary layer - as mentioned already before - all details of such flow fields can very well be predicted nowadays, and surprisingly enough, in acceptable agreement with experimental data. Some of the results obtained in investigations of this problems will be discussed in this paper later on. First, some results of numerical flow simulations in technical apparatus will be reported.

SIMULATION OF VORTICAL STRUCTURES IN A MODEL OF A PISTON ENGINE

Recently obtained cycle-resolved experimental data show that the flow inside the cylinder of a model of a piston engine with rectangular cross-section is dominated by large vortical structures during the intake and the beginning of the compression stroke [1]. The experiment, described in detail in the aforementioned reference, was exclusively designed to demonstrate the capability of numerical solutions of the Navier-Stokes equations for simulating strongly time dependent three-dimensional flows in technical apparatus. In the experiments the flow was visualized by using a Mach-Zehnder interferometer in co a high-speed camera with 6000 frames per second. The rectangular cross-section was necessary in order to come as close as possible to two-dimensional flow conditions. This requirement has to be satisfied for providing as small as possible density changes in the direction of the path of the light of the Mach-Zehnder interferometer. With this technique lines of constant density can be obtained as a function of time for the entire flow field during the intake and the compression stroke.

For the numerical simulation of the flow the Navier-Stokes equations were integrated numerically for time dependent, three-dimensional, compressible, laminar flows [2]. A turbulence model was not implemented in the solution for closure of the conservation equations, since existing closure models have not been validated to a sufficient degree of reliability for such flows. It was intended to simulate the flow directly without further approximation. An extended explicit Godunov-type method, second-order accurate in time and space was employed. One advantage of this method is that artificial damping terms need not be included in the solution. First results were computed with an algorithm for the simulation of two-dimensional flows on time dependent rectangular grids with 100 x 100 and 200 x 200 points. Although agreement with experimental data could be noted, marked deviation was found during the compression stroke. This observation suggested initiation of three-dimensional vortex structures with subsequent transition to turbulent flow. For that reason the method of solution was extended to three dimensions. Computations were carried out on time dependent grids with 100 x 100 x 50 points and with 120 x 100 x 50 points. For comparison with the experimental data the density contours were integrated in the direction of the path of the light of the Mach-Zehnder interferometer.

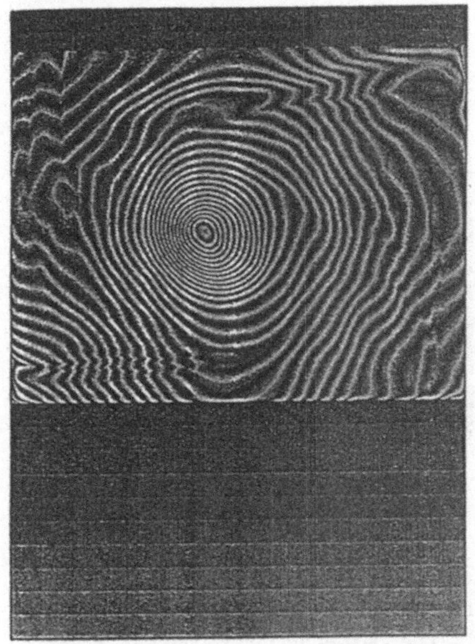

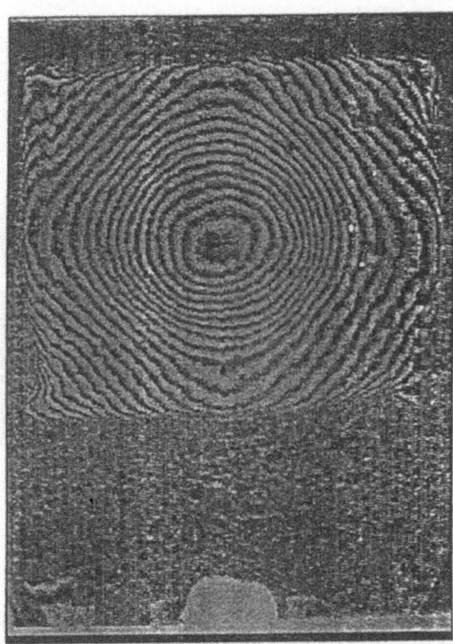

Fig. 1 Comparison of numerically simulated (left) and experimentally determined (right) density contours in a model of a piston engine with rectangular cross-section during the compression stroke for 1000 rpm at a crank angle of 270 °. Data were taken from [2].

The simulated density contours for three-dimensional flow agree remarkably well with the density contours obtained in the experiments. Fig. 1, taken from [3], shows a comparison of the integrated density contours with the experimental data for 1000 rpm during the compression stroke at a crank angle of 270 °. The head of the cylinder with rectangular cross-section is at the top of the picture, the piston at the bottom. The numerically simulated density contours are shown on the left, the experimentally determined contours on the right. The large vortex structure with its core in the center of the flow field can clearly be recognized. The predicted location, size, and density variation of the vortex agree with noticeable accuracy with the experimental data.

The deviation from two-dimensional flow during the compression stroke as predicted with the numerical solution of [2] can be seen in Fig. 2. Again computed density contours are shown, viewed in the direction of the motion of the piston for 1000 rpm during the compression stroke for crank angles of 270 °, 300 °, and 360 °. The density contours for 270 ° reveal already changes in the density in the axial direction of the main vortex, giving rise to axial flow in the core. Since that flow is not axially symmetric, the main vortex structure becomes three-dimensional in shape. With increasing compression, the perturbations become larger, and new small vortex structures are formed. Several of them can be identified in Fig. 2 on the density contours on the right for a crank angle of 360 °. The main vortex has not been destroyed by the compression, but its core is strongly deformed.

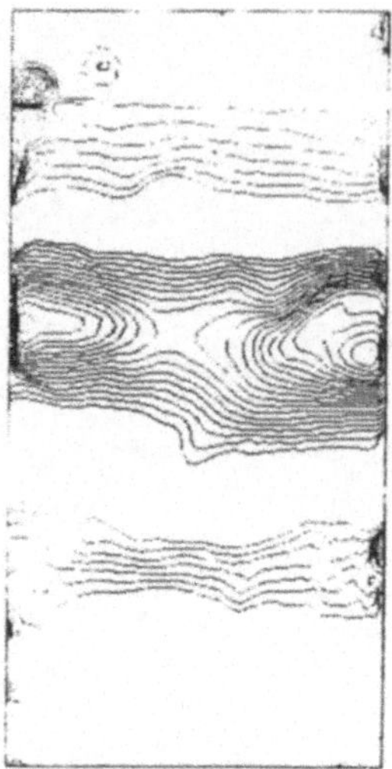 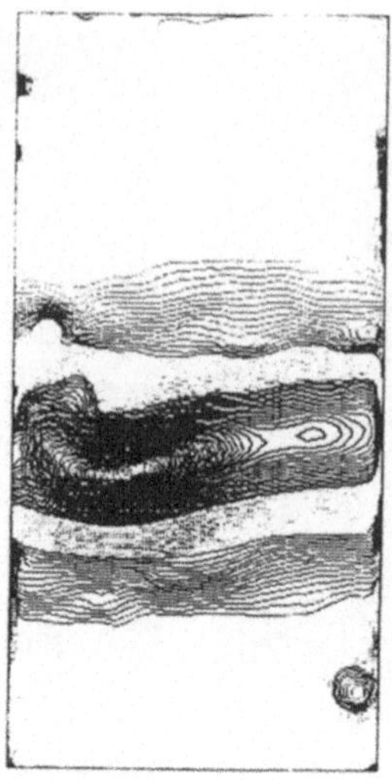 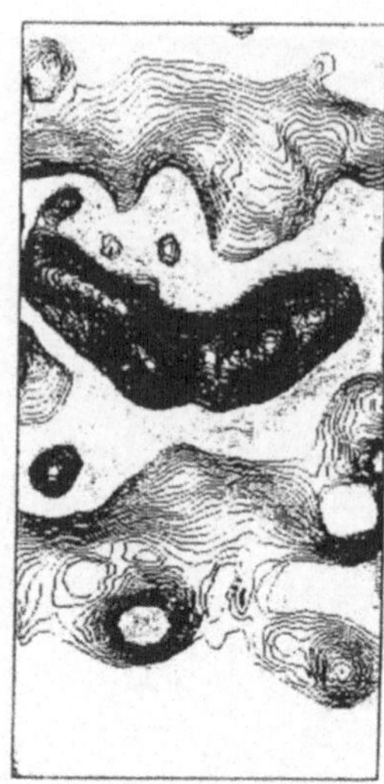

Fig. 2 Numerically simulated density contours in a model of a piston engine with rectangular cross-section during the compression stroke for 1000 rpm at crank angles 270 °, 300 °, and 360 °. The view is normal to the one in Fig. 1. Data were taken from [3].

It is not known yet, how much the initiation of three-dimensional flow in the Cylinder is influenced by truncation and other numerical errors. The experiments confirm, that the flow becomes turbulent near the top center, when the piston reaches its highest position. When the turbulence statistics are determined from the computed data, it is found that they are in qualitative agreement with LDA-measurements referenced in [3]. It is certainly understandable, that the small-scale vortical structures, which are of sub-grid size, cannot be resolved by direct simulation. In the flow investigated in [2] the smallest numerical length scales are considerably larger compared to the smallest physical length scales of the energy spectrum, and the spatial fluctuations of the density contours predicted in the computations are less pronounced than those observed in the experiments. But still these few remarks may suffice to show, what can be expected from future numerical investigations, if it is realized, that substantial increase in storage capacity and computational speed will allow a noticeable decrease in grid spacing.

SIMULATION OF THE BLOOD FLOW IN THE HUMAN CIRCULATORY SYSTEM

As an another example for numerically simulated internal flows a model for predicting the flow of the blood in the human circulatory system will briefly be described next. Although the prediction of biological flows in natural and artificial vessels, including rheological and chemical changes and interactions with the vessel walls, is a problem of enormous complexity. With appropriate simplifications, a solution, which includes only the time dependent variations in differential form, can be constructed in the frame of a system of ordinary nonlinear differential equations. In order to do so, the circulatory system is first decomposed into 15 elements, which in turn consist out of four active and eleven passive elements. They are depicted in Fig. 3. All passive elements are assumed to consist out of two parts; one being a rigid tube, establishing the connection to the next element, and the other is assumed to be a reservoir with elastic walls, for which the pressure-volume-relation is known.

The aorta, arteries, arterioles, capillaries, venules, veins, the venae cavae, the pulmonary arteries, arterioles, capillaries and veins constitute the passive elements in the model proposed (Fig. 3); they can only transport and store blood, but not activate any pumping mechanism. The blood flow is described by a set of Bernoulli's equation for un- steady flow, including a term approximating the pressure losses in each element, by another set of continuity equations, and the aforementioned pressure-volume-relations, which were constructed from physiological data.

The left and right ventricles and the left and right atriums are active elements. The pumping action of the ventricles is described by using Maxwell's model for simulating the contraction and the relaxation of the heart muscle, by converting the resulting forces into pressures, by incorporating an activation function to initiate the muscle action, and by assuming, that the actual shape of the ventricles can be approximated by a circular cylinder. The opening and closing of the four heart valves is described by varying the pressure-loss coefficient as a function of time, in the sense that the coefficient is small, if the valve is open, and large, if the valve is to be closed. If these assumptions and approximations are casted into mathematical form, two ordinary nonlinear differential equations are obtained; one for each ventricle, relating the time derivative of the blood pressure in the ventricle to the time derivative of its volume and the activation function. Finally, the system is closed by prescribing the pumping action of the atriums by activation functions. Thereby the system is closed. Altogether, a system of 32 nonlinear ordinary differential equations has to be integrated in time. This model was developed in [4], [5], and [6]. Latest applications can be found in [7].

Some of the results, obtained with the model described, are depicted in the following Figs. 4. and 5. In Fig. 4, the measured (left) and computed (right) pressure-time relations are shown for physiological conditions. First computations were carried out for heart rates of 70 beats per minute. The aim of these computations was to find out, how close the computation would simulate the working cycle of the heart. Fig. 4 shows, that the shape of the computed curves for the pressure-time relation are very similar to those obtained in experiments.

Similar agreement can be reported for the computed volume of the left ventricle as a function of time. The comparison with the measured values is shown in Fig. 5, where the experimental data are depicted on the left. The computed data on the right also show the volume of the right ventricle.

Further results are reported in [7]. There it is shown, that the model proposed can also simulate pressure and volume changes for higher heart rates than those mentioned here. Simulations were also carried out for some valvular abnormalities such as aortic insufficiencies and stenoses. The model was able to predict the changes in the pressure-time and volume-time relations. It can be expected that investigations similar to the one, briefly described here, will find applications to a much larger extent than was possible until now, as, for example, for the purpose of reducing animal experiments.

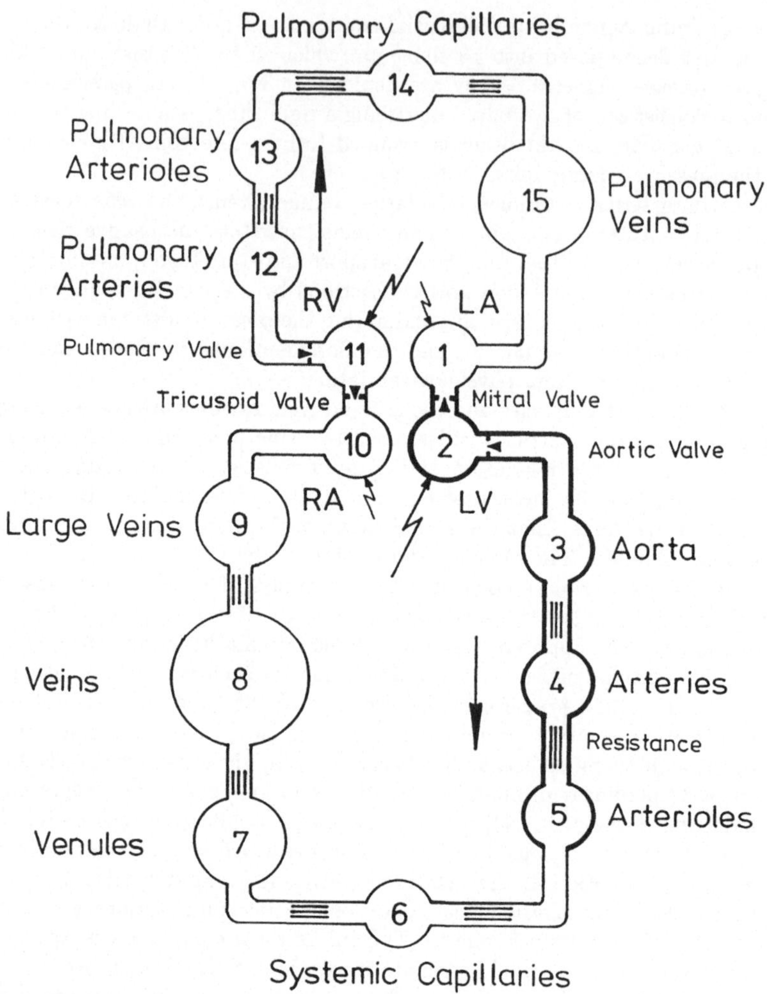

Fig. 3 Decomposition of the circulatory system into active and passive elements for prediction of blood flow. The active elements, the left atrium (1) and ventricle (2), and the right atrium (10) and ventricle (11) exert pumping functions; the passive elements, the aorta (3), arteries (4), arterioles (5), systemic capillaries (6), venules (7), veins (8), large veins (9), the pulmonary arteries (12), arterioles (13), pulmonary capillaries (14), and pulmonary veins (15) exert transporting and storing functions.

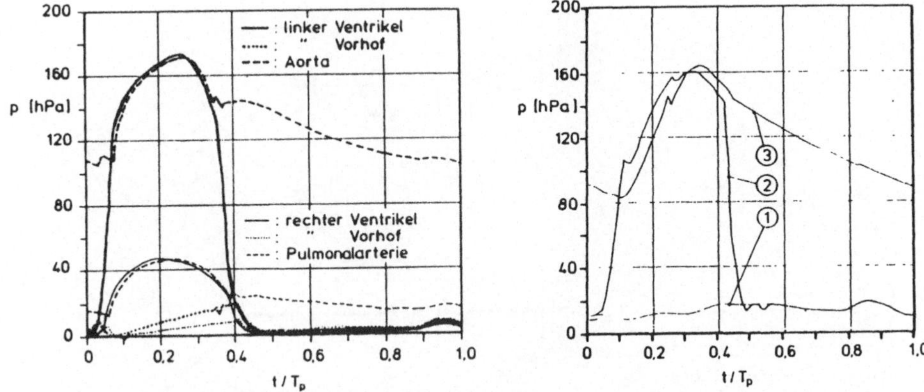

Fig. 4 Measured (left) and computed (right) pressure-time relations for physiological conditions. The measured values are shown for the left and right ventricle, the left and right atrium, for the aorta, and for the pulmonary arteries. The computed values are for the left ventricle, atrium and aorta. The heart rate is 70 beats per minute. Note that the computed curves are very similar in shape compared to those obtained in the experiments. The data were taken from [6].

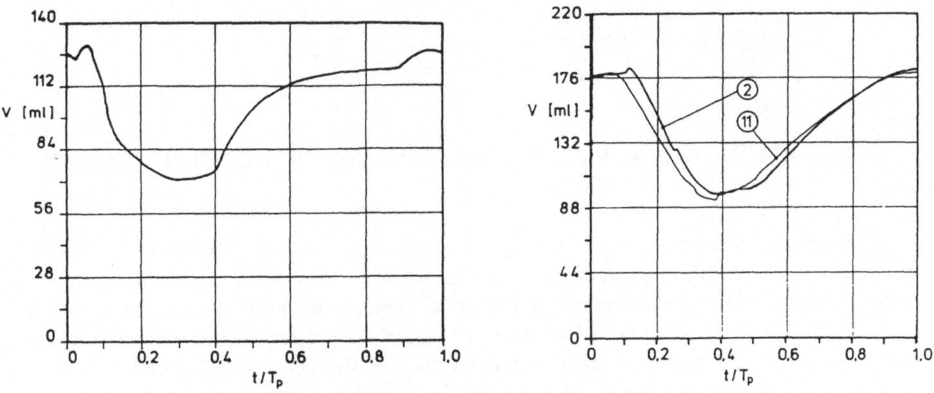

Fig. 5 Measured (left) and computed (right) volume of the left ventricle for physiological conditions. The figure on the right also shows the computed volume of the right ventricle. The heart rate is 70 beats per minute. The shape of the curves obtained in the computation is very similar to the shape of the curve obtained in the experiments. The data were taken from [6].

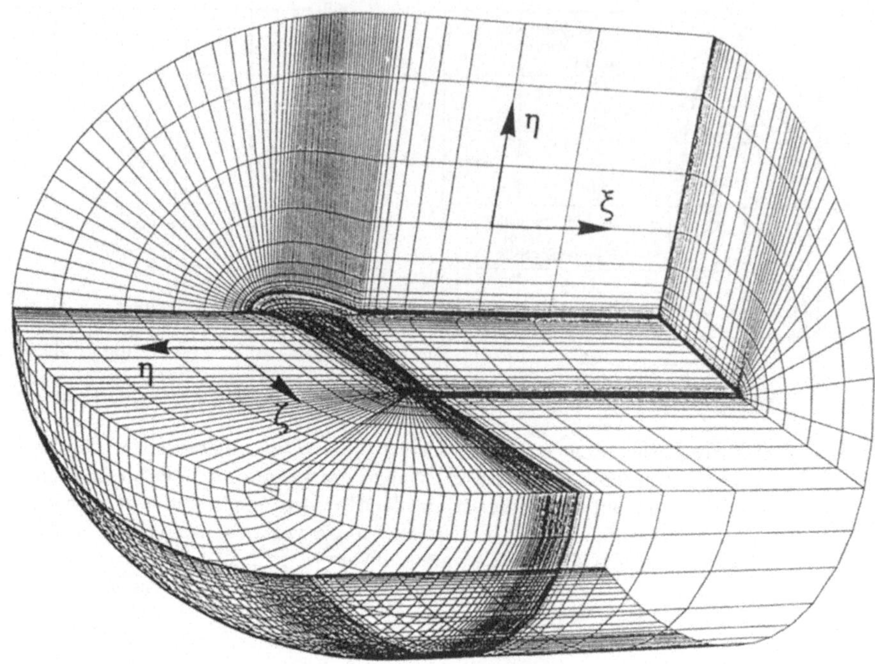

Fig. 6 Grid arrangement used in [8] for computation of transonic flow around a wing, here the DFVLR - F 5 wing.

SIMULATION OF FLOW AROUND AERODYNAMIC CONFIGURATIONS

This section aims at demonstrating the application of numerical techniques to the solution of aerodynamic problems. It is well known, that work in aerodynamics has always pioneered the development of numerical techniques, and it seems, that this will continue in the years to come. Most likely the main reason for continuously stimulating new research on numerical integration techniques and further development of computing machines is the extreme accuracy required for aerodynamic computations. Although the lift of aerodynamic configurations can be predicted with sufficient accuracy, the accuracy of the computed drag leaves much to be desired. Here again, one hits upon the unsolved problem of predicting transition, the turbulent flow structure in both, the boundary layer and the wake of the wing. For transonic, supersonic, and hypersonic flows the complexity of the flow is even enlarged by the occurrence of shock waves, interacting with the boundary layer, and by real gas effects at high speeds, affecting the entire flow field substantially and increasing the necessary computational effort often beyond the capability of presently available solution techniques.

Nevertheless, substantial progress has been made during recent years for the entire speed regime. The most noticeable advancement is, perhaps, that investigations of threedimensional flows have almost become commonplace. Solution techniques re-

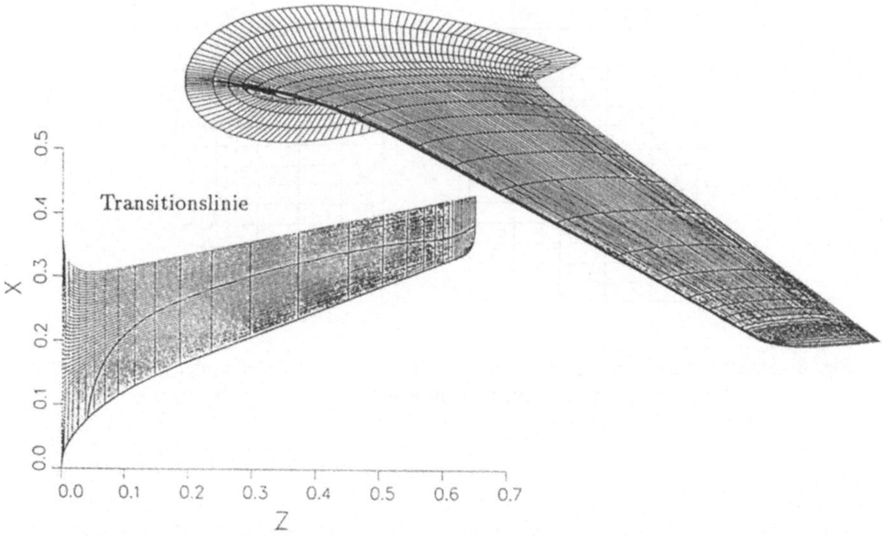

Fig. 7 Surface grid arrangement for computation of transonic flow around a wing, here the DFVLR - F 5 wing, taken from [8].

stricted to twodimensional problems have little hope for survival. In transonic aerodynamics, fu investigations will mainly concentrate on improving drag predictions and on analyzing near- and far-wake problems. Solutions for these types of problems require refined grid-arrangement strategies, in order not to miss the major events in the flow. Fig. 6 shows a grid arrangement, typical for the computation of transonic flow around a wing, in this case the DFVLR - F 5 wing [8].

It is seen, that the wake contains relatively few grid points; hence, the vortic al structures in the wake can barely be caught in the computation. Substantial incr ease in the number of grid points will be necessary, if the roll-up process of tip vortices or other vortices, originating, for example, from the root of the wing or from extended f laps. The accuracy of the drag prediction will - in addition to the modelling of the trans ition process and of the turbulent structure of the flow in the boundary layer - also depend o n the surface grid spacing. Since it can be expected that large-eddy simulation of the flow ar ound a wing will be possible in the next few years, surface grid refinement will also be nec essary for the wing. A presently used surface grid is shown in Fig. 7 for the DFVLR F 5 wing, taken from [8].

As pointed out before the lift can be computed with sufficient accuracy. This can be seen in Fig. 8, where the pressure distribution measured on the upper surface of the DFVLR - F 5 wing is plotted for an arbitrary section z/s = 0,492. The measurements are indicated by the symbols. The curve in the left part of Fig. 8 is obtained with the method of [9].

The good agreement of computed and measured pressure distributions in the chordwise direction can also be noted in the spanwise direction. In Fig. 9, the isobars

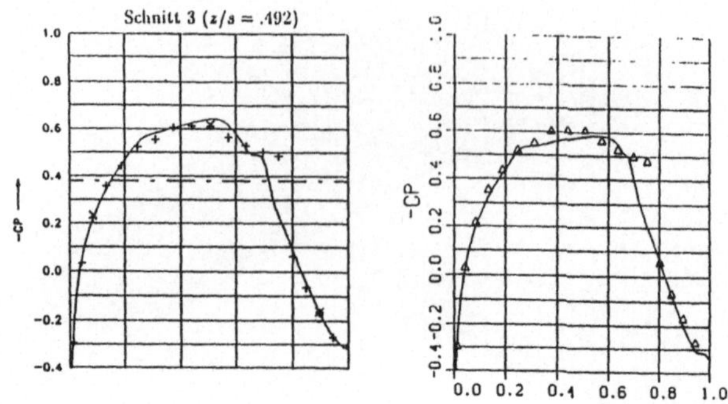

Fig. 8 Measured and computed pressure distribution on the upper side of the DFVLR - F 5 wing. Computed values shown in the left diagram are taken from [9], those in the right from [8].

computed with the two methods described in [8] and [9] are compared with each other for the upper side of the wing.

Good agreement can also be obtained for supersonic and hypersonic flow conditions. Extensive flow computations were recently carried out for the hypersonic research configuration "ELAC I" of the Collaborative Research Center (SFB 253) "Fundamentals of Design of Aerospace Planes", sponsored by the German Research Association (DFG) at the Aachen University of Technology since 1989. In [10], flow computations were reported for freestream Mach numbers of 2 and 7.9. The comparison with experimental data of [11] and [12] showed persuasive agreement, although all computations were carried out for laminar flow conditions. Also the grid was not refined, since the computations were not aimed at obtaining best agreement with the experimental data, but rather to provide reference computations.

The research configuration "ELAC I" is a thick delta wing with a rounded leading edge. Its cross-section is formed by two half ellipses of equal large, but different small axes. The thickness increases linearly to a maximum value, and decreases again linearly to the trailing edge. Two controls are mounted near the trailing edge. The configuration is not equipped yet with a propulsion system. Its integration will be carried out for the follow-up configuration "ELAC II". The shape of "ELAC I" and the surface grid used in the computations is shown in Fig. 10.

Fig. 11 shows a comparison of computed and measured pressures for free-stream Mach numbers $Ma_\infty = 2.0$ and $Ma_\infty = 7.9$. It is seen, that the computed values follow the overall tendency of the measured values, although the grid was not refined, and transition to turbulent flow was not taken into account in the computations. For $Ma_\infty = 7.9$, experimental data exist so far only for the upper side of the configuration. Because of the slenderness of the model, pressure measurements can be obtained only on either the lower or the upper side. The pressures on the lower side will be determined in future experiments.

Similar agreement is also obtained for pressure distributions in different sections and for different angles of attack. Here again, the comparison of computed and experimental data shows that predictions can be relied on, provided, the influence of transition and

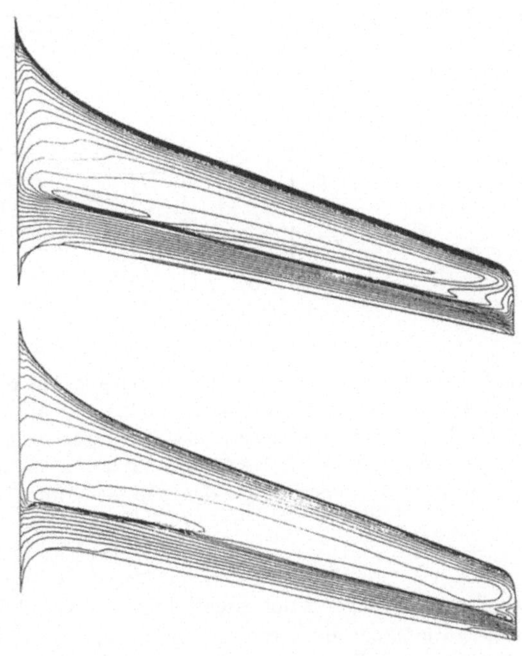

Fig. 9 Isobars computed for the upper side of the DFVLR - F 5 wing with the methods described in [8] (lower part of Fig. 9), and with the one described in [9] (upper part).

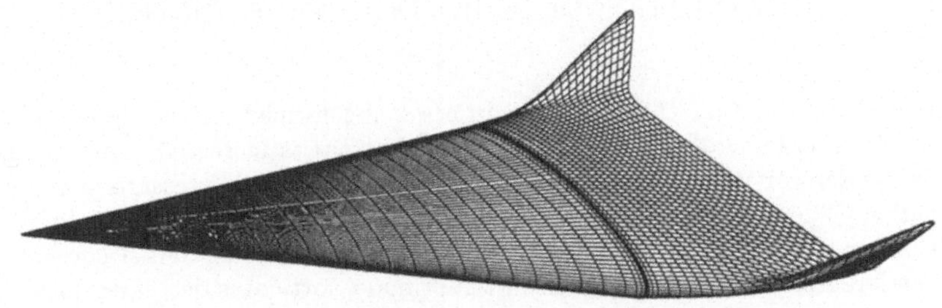

Fig. 10 Surface grid used for computation of the flow around the research configuration "ELAC I" of the Collaborative Research Center (SFB 253) "Fundamentals of Design of Aerospace Planes", sponsored by the German Research Association (DFG) at the RWTH Aachen since 1989. Taken from [10].

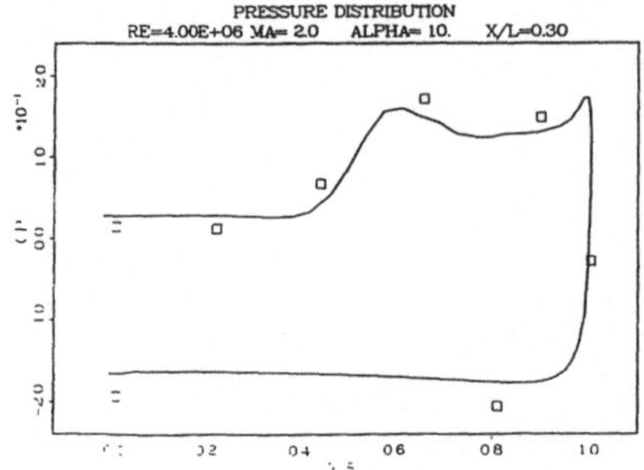

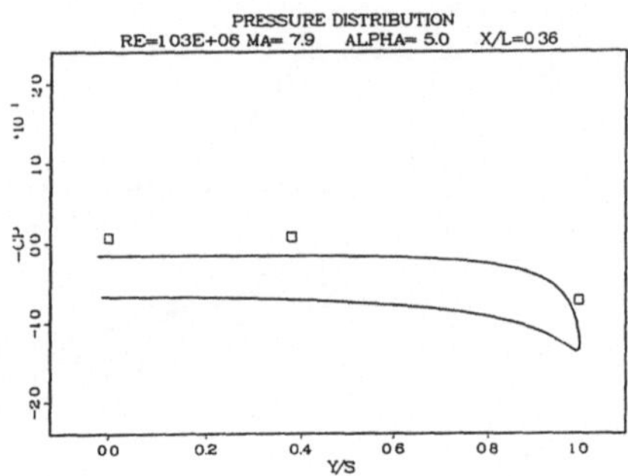

Fig. 11 Computed and measured pressure distributions for the research configuration "ELAC I". The freestream Mach numbers are $Ma_\infty = 2.0$ and $Ma_\infty = 7.9$. For $Ma_\infty = 7.9$ experimental data are available so far for the upper side of the configuration. Corresponding values for the lower side will be determined in future experiments. Data are taken from [10].

turbulent flow in the boundary layer is not essential for the overall flow pattern. Further details of these investigations can be found in [10] and other articles published in Vol. 17 (1993) of the Z. Flugwiss. Weltraumforsch. The capability of numerical methods for analyzing hypersonic flows in nozzles and around space planes was recently persuasively demonstrated in [13]. Those results leave no doubt, that development of modern technology is not possible without involving numerical simulation tools.

SIMULATION OF FLOWS IN SLENDER VORTICES

It is well known, that the primary vortices being formed on the upper side of delta wings tend to break down, when the angle of attack is increased. One mechanism to force slender vortices to break down is the increase of static pressure in the main flow direction. Thereby, the pressure near the axis of the primary vortices is increased, and the pressure gradient in the radial direction becomes smaller. As a consequence, the circumferential velocity component of the primary vortices is decreased locally, leading eventually to stagnant flow and destruction of the vortex core. This process, which is of great importance for the flight characteristics of delta wings, particular for low-speed flight, with a high angle of attack, was studied in great detail in numerical and experimental investigations. The analysis of this problem is extremely complicated by the observation that the flow may undergo transition from one flow mode to another, generally referred to as bubble- and spiral-type breakdown. The flow is three-dimensional and unsteady, such that a simplification of the conservation equations cannot lead to success. Fig. 12 shows an example of the simulation of bubble-type breakdown.

Continuing the simulation shows that the core of the vortex is distorted into spiral-breakdown, Fig. 13. Since the exact boundary conditions of the experiments could not be determined, quantitative comparison is not possible yet.

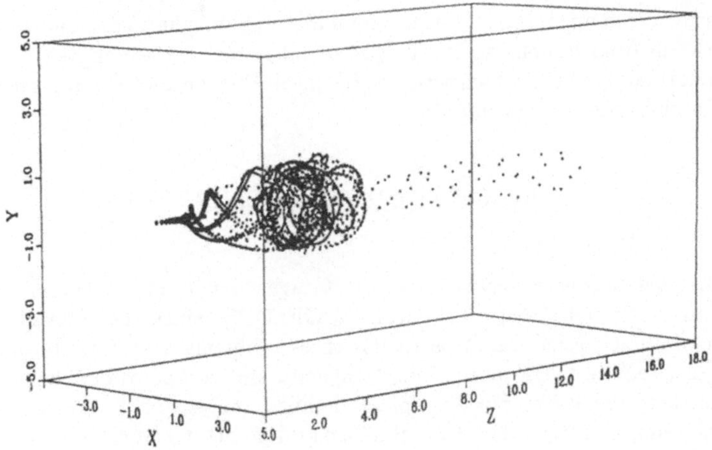

Fig. 12 Computed streaklines for bubble-type breakdown of an isolated slender vortex shortly before transition to spiral-type breakdown. Picture taken from [14].

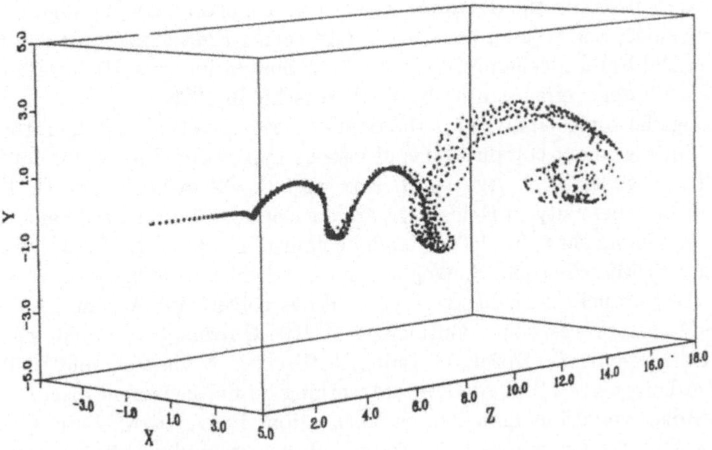

Fig. 13 Computed streaklines for spiral-type breakdown of an isolated slender vortex. After [14].

Presently, numerical investigations are under way, in which conditions to be satisfied for transition from bubble- to spiral-type to take place are attempted to be detected. This is a formidable task, which seems to be unsolvable without making use of numerical tools of high-performance computing.

REMARKS ON PARALLELIZATION

Presently available vector machines offer a storage capacity of approximately 1 GBYTE and a computational speed of about 1 - 5 GFLOPS. Since the solutions of complex problems in fluid mechanics, in particular those problems, which are termed as "grand challenges", require much more storage capacity and computational speed; both have to be increased markedly. For example, a detailed analysis of the transonic flow about wings will require of the order of 150 millions grid points in comparison of a few millions for problems being currently investigated. In order to attain the required speed, parallel machines are used in increasing number. They pose new problems in the sense that the investigated flow field has to be split into a number of sub-domains. They are assigned to the processors, such that therefore communication between the various processors must be established.

Since these problems, generally referred to as domain decomposition and load balancing problems are relatively new, roughly one third of the investigations of the DFG-sponsored Priority Research Program "Flow Simulation with High-Performance Computers" is focussed on the development of new, or porting existing algorithms to parallelized versions, and to gain experience with them. First results are reported in Notes on Numerical Fluid Mechanics, Vol. 38 "Flow Simulation with High-performance Computers I", Vieweg Verlag, edited by E. H. Hirschel in 1993.

Among the topics analyzed is the domain decomposition and operator splitting for parallel finite element computations of viscous hypersonic flow taking into account the effect of real gases by J. Agyris, H. Friz, and F. Off of Stuttgart; G. Bader and E. Gehre of the University in Heidelberg are developing a parallelized version of the solution for the flame sheet model for laminar flames; A. Bode, M. Lenke of the Technical University in München and S. Wagner and T. Michl of Stuttgart University are working on the parallelization of an implicit Euler solver for Alliant FX/280 and Intel IPSC/860 multiprocessors; J. Burmeister of Kiel University is developing time-parallel multi-grid methods; O. Dorok, G. Lube, U. Risch, F. Schieweck, and L. Tobiska of the Technical University of Magdeburg are working on finite element discretizations of the Navier-Stokes equations and their parallelization; F. Durst, M. Peric M. Schäfer, and E. Schreck of the University Erlangen-Nürnberg are implementing a grid partitioning technique for the parallelization of a multi-grid finite volume algorithm for the prediction of two-dimensional laminar flow in complex geometries; P.-W. Gräber and Th. Müller of the University of Technology Dresden are employing transputer networks in the simulation of migration processes in the soil and groundwater zone; M. Griebel, W. Huber, and C. Zenger of the Technical University München are developing a fast Poisson solver for turbulence simulation on parallel computers using sparse grids; W. Juling and K. Kremer are reporting on their experience with a transputer-based parallel system developed by the Parsytec GmbH; F. Lohmeyer and O. Vornberger of the University of Osnabrück implemented a parallel algorithm, which is based on an explicit finite element scheme to solve the Euler and Navier-Stokes equations; M. Meinke and E. Ortner of the RWTH Aachen implemented explicit Navier-Stokes solvers on massively parallel systems; finally, S. Pokorny, M. Faden, and K. Engel of the DLR in Köln are developing a parallelized simulation system to compute three-dimensional unsteady flow in turbo machines.

Details of the investigations may be found in Vol. 38 of the Notes on Numerical Fluid Mechanics. In closing it is noted, that information about bench mark results, scalability, speed up, efficiency, and other aspects of problems relating to the use of parallel and massively parallel systems is discussed therein. Most likely, future work will include development of parallel algorithms to a much larger extent than previously expected.

CONCLUDING REMARKS

Several examples of applications of high-performance computing in fluid mechanics were reported. These include the simulation of compressible, time dependent, and threedimensional flow in a cylinder of a model engine with rectangular cross-section; of time dependent pressure and volume variation of the blood flow in the human circulatory system; of transonic, supersonic, and hypersonic flows around wings; of flows in slender vortices. A brief overview was given of the development of new parallel algorithms and porting existing codes to parallel versions.

The examples described in the text clearly demonstrate, that future work in fluid mechanics will strongly engage high-performance computing. Improvements of the algorithms, the storage capacity, and the computational speed will soon allow numerical simulations on grids with 150 millions and more points. It was shown, that validation of data by experiments is - and will be - of extreme importance.

What could not be documented in the text is the fairly new experience with time dependent three-dimensional numerical flow visualizations. They will soon become an indispensable tool in the analysis of complex flow phenomena.

REFERENCES

[1] Jeschke, M.: "Zyklusaufgelöste Dichtefelder in einem geschleppten Modellmotor mit quaderförmigen Innenraum", Diss. Aerodyn. Inst., RWTH Aachen, (1992).

[2] Klöker, J.: "Numerische Simulation einer dreidimensionalen, kompressiblen, reibungsbehafteten Strömung im Zylinder eines Modellmotors", Diss. Aertodyn. Inst., RWTH Aachen, (1992).

[3] Klöker, J., Krause, E., Kuwahara, K.: "Vortical Structures and Turbulent Phenomena in a Piston-Engine Model", Lecture Notes in Physics, 414, Proceedings of the Thirteenth International Conference on Numerical Methods in Fluid Dynamics, Rome, Italy, 6 - 10 July 1992, M. Napolitano and F. Sabetta (Eds.), SpringerVerlag, (1993), pp. 165 - 169.

[4] Steinbach, B.: "Simulation der Ventrikeldynamik bei Klappenersatz und Pumpunterstützung", Diss. Aerodyn. Inst., RWTH Aachen, (1981).

[5] Bialonski, W.: "Modellstudie zur Entlastung des linken Herzens", Diss. Aerodyn. Inst., RWTH Aachen, (1987).

[6] Langner, F.: "Numerische Simulation der Blutströmung im menschlichen Kreislauf", Abh. Aerodyn. Inst., RWTH Aachen, 30, pp. 8 - 19, (1990).

[7] Zácek, M., Krause, E. "Simulation der Blutströmung im menschlichen Kreislauf bei physiologischen und pathologisch veränderten Bedingungen", Abh. Aerodyn. Inst., RWTH Aachen, 31, Veröffentl. in Vorbereitung; (Ende 1993).

[8] Seider, G.: "Numerische Untersuchung Transsonischer Strömungen", Diss. Aerodyn Inst., RWTH Aachen, (1991).

[9] Schwamborn, D.: "Simulation of the DFVLR - F 5 Wing Experiment Using a Block-Structured Explicit Navier-Stokes Method", Notes on Numerical Fluid Mechanics, Vol. 22, W. Kordulla (Ed.), Vieweg Verlag, (1988), pp. 244 - 268.

[10] Hänel, D., Henze, A., Krause, E.: "Supersonic and Hypersonic Flow Computations for the Research Configuration ELAC I and Comparison to Experimental Data", Z. Flugwiss. Weltraumforsch., 17, pp. 90 -98, (1993).

[11] Jessen, C., Vetter, M., Grönig, H.: "Experimental Studies in the Aachen Hypersonic Shock Tunnel", Z. Flugwiss. Weltraumforsch., 17, pp. 73 - 81, (1993).

[12] Limberg, W., Stromberg, A.: "Pressure Measurements at Supersonic Speeds on the Research Configuration ELAC I", Z. Flugwiss. Weltraumforsch., pp. 82 - 89.

[13] Weiland, C., Schröder, W., Menne, S.: "An Extended Insight into Hypersonic Flow Phenomena Using Numerical Methods", Computers Fluids, Vol. 22, No 4/5, pp. 407 - 426, (1993).

[14] Breuer, M., Hänel, D.: "A Dual Time-Stepping Method for 3-D, Viscous, Incompressible Vortex Flows", Computers Fluids, Vol. 22, No 4/5, pp. 467 - 484, (1993).

PROCESS SIMULATION FOR THE SEMICONDUCTOR INDUSTRY

J. Lorenz[1], F. Durst[2], H. Ryssel[31]
Bayerischer Forschungsverbund für Technisch-Wissenschaftliches Hochleistungsrechnen
[1] Fraunhofer-Institut für Integrierte Schaltungen, Bereich Bauelementetechnologie
Schottkystrasse 10, D-91058 Erlangen, Germany
[2] Lehrstuhl für Strömungsmechanik, Universität Erlangen-Nürnberg
Cauerstrasse 4, D-91058 Erlangen, Germany
[3] Lehrstuhl für Elektronische Bauelemente, Universität Erlangen-Nürnberg
Cauerstrasse 6, D-91058 Erlangen, Germany

SUMMARY

A brief overview of approaches used and problems encountered in the simulation of the semiconductor fabrication process steps ion implantation, diffusion, oxidation, and layer deposition is given. Some recent improvements obtained at Erlangen in the development of physical models are described. Problems occuring for two- and especially for three-dimensional simulations are outlined. Their relationships to high-performance computing are sketched, and approaches for their solution are developed. Model and program requirements are discussed from the viewpoint of the requirements of industrial application. Furthermore, the importance of equipment simulation is discussed and an example for this strongly developing field is given.

INTRODUCTION

During the last couple of years, the use of simulation programs has become indispensible to support the development and optimization of microelectronic devices and integrated circuits. Various kinds of simulation programs have been used e.g. for IC layout design, device, and process simulation. Whereas the main attention was originally placed on design issues, the microscopic description of the processes occuring during device fabrication is becoming more and more important. The reason for this development is that with shrinking device dimensions the relative differences between the theoretical feature sizes used in IC layout, e.g. mask window spacings, and their real values in or on silicon after device fabrication are considerably growing. The microscopic description includes both the classical field of process simulation which describes the modification of geometry and dopant distribution in a device during the fabrication process and the field of equipment simulation in which both the nominal process influence on the wafer and its inhomogeneities are calculated from the machine parameters.

The use of process simulation allows for a reduction of the time and costs spent in the development of devices and circuits: A considerable part of runs which are being used for the experimental optimization are replaced by simulation runs. In future, this trend is expected to increase, because the capabilities of process simulation tools are considerably growing and the costs for computer use are decreasing, in contrast to the costs to carry out experiments. Furthermore, for deep-submicron devices it is very difficult to measure key parameters such as the lateral shape of a dopant profile below the gate of a transistor. This demand is driving the development of more accurate physical models, more general simulation programs, and the transition from one- and two-dimensional simulation towards multidimensional simulation systems which are capable of simulating critical aspects in three dimensions.

ION IMPLANTATION

In order to be able to simulate the electrical behavior using a device simulation program it is necessary to start from reliable information on the spatial distribution of the dopant atoms in the device in question. Inhomogeneous dopant distributions are introduced into the device by ion implantation or deposition of doped layers, and are affected by subsequent high temperature steps. With shrinking device dimensions it is necessary to reduce dopant diffusion in device fabrication in order to obtain sufficiently shallow junctions. In consequence, dopant profiles in critical parts of the device are often mainly controlled by the ion implantation step which introduces the dopant, and are not considerably modified in shape in subsequent annealing steps. For submicron devices it is extremely difficult to measure two-dimensional dopant profiles with sufficient accuracy. Process models can be developed and calibrated using specific test structures which are easier to measure. Therefore it is necessary to follow this approach and to develop methods to accurately model the dopant distributions resulting from ion implantation.

Three different approaches are used for the simulation of ion implantation: In Monte-Carlo simulations, many individual trajectories of ions are traced until the ion energy is low enough to consider the ion to be stopped. Programs are existing for amorphous targets, e.g. [1], or for crystalline targets, e.g. [2]. The main differences between these kinds of Monte-Carlo programs are the different use of random numbers to determine the impact parameters for the scattering processes at the substrate. So-called "dynamical" Monte-Carlo programs like [3] also consider the modification of the implanted layer due to the generation of defects and the destruction of the crystalline structure. The stopping and scattering of the implanted ions can in principle be described by physically based models which only need some special improvements. However, a major problem for the development of Monte-Carlo programs is to check in which layer of a multilayer target the ion is located at a point of time. Because the ion path consists of a large number of polygon segments, efficient algorithms are necessary to avoid too large CPU times for each single trajectory. Considerable progress has been obtained on this problem elsewhere [4] by a concept which uses quadtrees in 2D, or octrees in 3D, for the location of the ions. Anyhow, because some 100 000 ion trajectories must be traced in order to get a 2D profile with a reasonable statistics, 2D Monte Carlo simulations require CPU times of hours to days on current high-performance workstations. This still prevents the standard use of this method in industrial applications.

The second approach for the simulation of ion implantation is the solution of the Boltzmann transport equation [5]. This method, however, involves a considerable mathematical complexity. Even for a 1D simulation, the phase space to be considered has a dimension of three, because the energy and also the projection of the velocity onto the direction of interest must be considered. This leads to a very large number of discretization points and large CPU requirements not only in terms of computation time but also in terms of memory. For this reason, this method is hardly used for the multidimensional simulation. However, the problem is extremely simplified by expanding the solution into Legendre-polynominals and then calculating moments of the equation [6]. In this way, tables of vertical, lateral, and mixed range moments for implantation into homogeneous amorphous targets can be calculated e.g. for energies between 1 keV and 10 MeV within some hours CPU time, depending on the workstation used. These tables are then stored and can afterwards be used to reconstruct the dopant distributions with the analytical approaches outlined below. This approach has been implemented in the program RAMM [6].

Analytical models for the description of ion implantation have been developed since nearly two decades [7,8] and have in the meantime achieved a high level of maturity. Because the slowing down of ions in a target is a statistical process it is not surprising that distribution functions known from statistics have been complemented successfully with physical reasoning and measured or theoretically calculated parameters in order to arrive with an appropriate description of implanted ion distributions for a wide variety of applications. In general the benefit of this approach is that the CPU time needed for a 2D simulation is about one to a few minutes. This makes the analytical approach very attractive for industrial applications.

The analytical approach involves various levels of sophistication. In summary, the vertical dopant distribution in a single layer is mostly described by Pearson distributions [7]. Crystalline targets require the use of Pearson IV distributions, whereas for the implantation into amorphous targets mostly Pearson VI or I distributions are appropriate [9]. Dose-dependent residual channeling can either be considered by weighted superposition of Pearson distributions for the amorphous and the crystalline component, respectively [10], or by using dose-dependent range moments. Implantation through masking layers is appropriately described by the Numerical Range Scaling Model [11,12] which modifies the projected range and the projected range straggling in the silicon to take into account the different electronic straggling and nuclear scattering in the masking layers.

The simulation of the lateral shape of the implanted ion profiles is receiving increasing attention. In submicron devices, the lateral profiles are very important for electrical behavior and long-term stability. However, there are very limited possibilities to measure these profiles in 2D. Advanced analytical models have been developed, using comparisons with Monte-Carlo simulations, and are being evaluated against the few 2D measurements becoming available. The basic principle of current lateral models is a convolution of the vertical model used with either a Gaussian or, more general, a Pearson II or Pearson VII lateral distribution, depending on the values of the lateral range moments. Recent developments made at Erlangen include models for the depth-dependent lateral spread (second centered lateral moment) in single layers and in multilayer structures, and for the depth-dependent lateral kurtosis (forth centered lateral moment, normalized by the forth power of the lateral spread). For the depth-dependent lateral spread $\triangle R_{pl}(x)$, a mixed

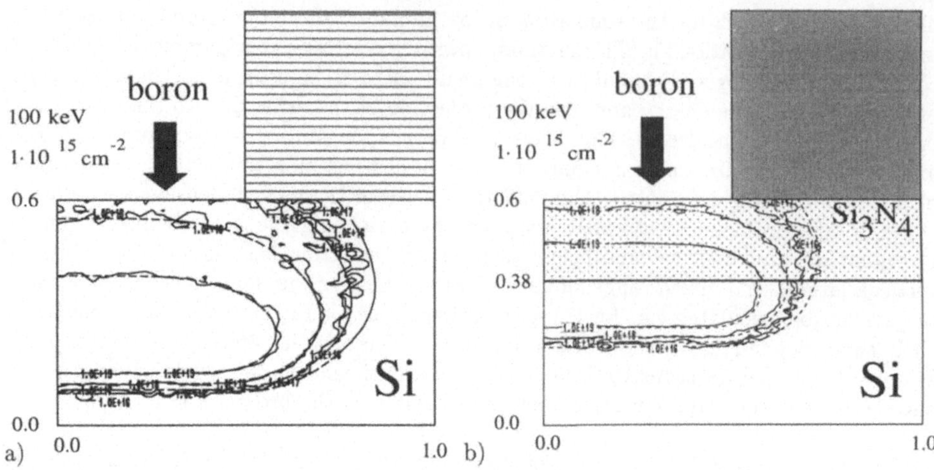

Fig. 1: Comparison between analytical model (broken lines) and Monte-Carlo simulation for the implantation of boron with an energy of 100 keV near a vertical mask edge: a) Implantation through mask window into bare silicon. b): Implantation through mask window and 0.2μm nitride layer on silicon

parabolic-exponential approach has been developed which agrees well with Monte-Carlo data [13]:

$$\triangle R_{pl}(x) \;=\; \triangle R_{pl} \cdot \left\{ \begin{array}{ll} \sqrt{1+a+b\cdot z-a\cdot z^2} & : z \leq 0 \\ exp(A+B\cdot z+C\cdot z^2) & : z > 0\,. \end{array} \right. \tag{1}$$

with the mean value for the lateral spread $\triangle R_{pl}$, the projected range R_p, the projected range straggling $\triangle R_p$, and the normalized depth $z = (x - Rp)/\triangle R_p$. For ions which are heavier than the target material, the backscattering behavior is different. In that case, the exponential expression must be used between $x = 0$ and $x = R_p$, whereas the parabolic one has to be used beyond R_p. The parameters a and b can be expressed in terms of the vertical, lateral, and mixed moments calculated by RAMM, whereas the parameters A, B, and C can be related to a and b via Taylor expansion of the exponential in eq. (1). The example given in Fig. 1a illustrates that this model gives good agreement with Monte-Carlo simulations. However, in order to be valuable for industrial applications, the model must also be applicable for the implantation through non-ideal mask edges, i.e. mask edges which are not oriented parallel to the implantation beam. The multilayer model mentioned above adds to the depth-dependent lateral spread in the substrate a positive or negative shift which takes into account the different scattering behavior of the mask. An example is given in Fig. 1b which is the same situation as in Fig. 1a except that in this case the implantation is performed through a layer of 200μ m Si3N4. The comparison of the two figures reveals that not only the depth of the implanted profile is reduced by the nitride layer but also its lateral extent in the silicon. This effect is important for current semiconductor devices like LATID (Large Angle Tilted Implant Device). A more detailed description of this model is being published elsewhere [14].

For the multilayer models, the distances an ion would travel on a straight line through all layers on top of the point of interest must be known. These are difficult to obtain within reasonable CPU times in case of highly nonplanar structures, e.g. overhanging mask edges or the "standing wave" patterns observed in optical lithography. Furthermore, with a few exceptions an ion implantation step drastically changes the dopant distribution in the wafer, both in case of an inhomogeneous distribution alredy existing before the implantation step or not. Therefore, in order to be able to sufficiently resolve the implanted dopant profile with a reasonable number of discretization points it is mandatory to perform a mesh adaptation procedure during the simulation of ion implantation. These steps are generally followed by diffusion steps. Therefore, it is necessary that the mesh used and adapted in ion implantation is compatible with the one used in dopant diffusion: The mesh should only be modified in parallel with the change of the dopant profile, not already by simply starting the diffusion step. Within the current activities an approach has been developed which combines the efficient calculation of the straight-line distances mentioned above with the generation and adaptation of a mesh compatible with dopant diffusion [15,16].

HIGH TEMPERATURE STEPS

After ion implantation, the wafers need to be annealed at high temperatures in order to remove the damage generated and to activate the dopant atoms introduced. Furthermore, high temperature steps are also used for intentional modification of the dopant profile, e.g. the formation of wells, and for the growth of oxide layers needed for isolation purposes or as masks for subsequent process steps. Therefore, in process simulation it is mandatory to appropriately model the diffusion and activation of dopants in these high temparature steps.

A lot of physical effects need to be considered in dopant diffusion. This includes the different charge states of the dopants, the influence of the electrical field caused by the dopants, the generation and diffusion of point defects and their influence on the diffusion of the dopant atoms, dynamical clustering and precipitation of the dopants, segregation of the dopants at material interfaces or at grain boundaries in polycrystalline layers, diffusion in grains and along grain boundaries in polysilicon, and influences of the environment on the diffusion of dopants in oxides. For the formation of shallow junctions, Rapid Thermal Annealing (RTA) is frequently used after ion implantation. In such processes, transient diffusion occurs due to the interaction between the dopant atoms implanted and the point defects generated during the implantation step. To model this, the system of coupled partial differential equations to be solved includes at least one equation for each dopant species, one for vacancies, and one for interstitial atoms, plus one reaction equation for the clustering of the dopant. An example for such a model where the driving forces for dopant diffusion are not only the dopant gradient and the electrical field but also the gradients of the vacancy and interstitial concentrations was presented by Orlowski [17]. More rigorous physical models require the solution of partial differential equations for pairs of dopants and point defects in various charge states and, therefore,

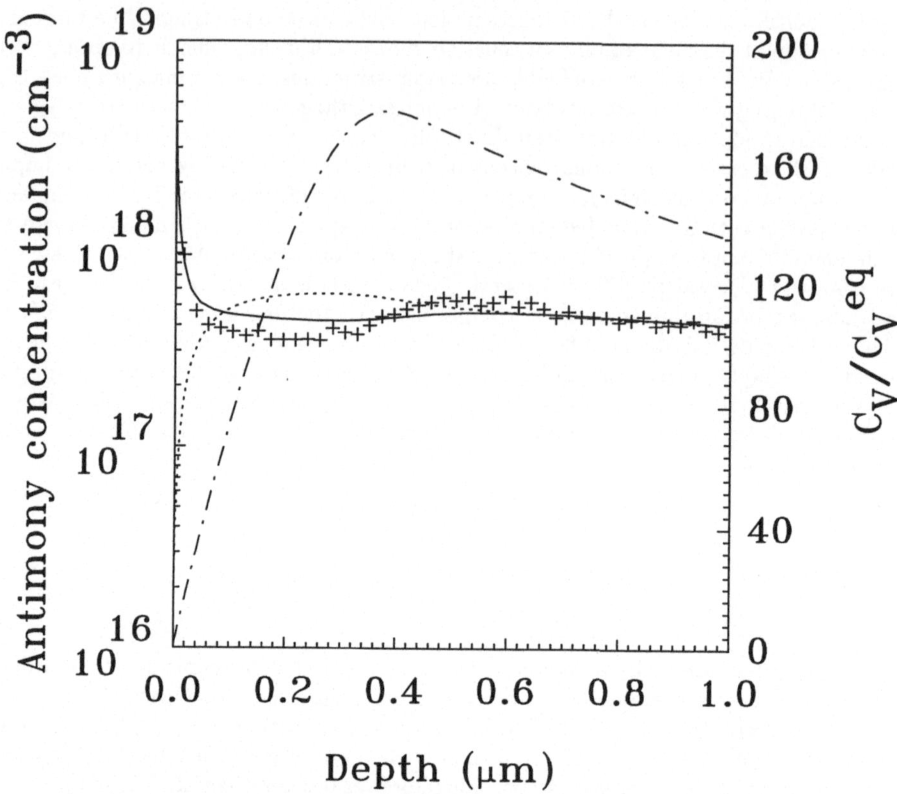

Fig. 2: Comparison between results from the pair diffusion model (drawn line), the model of Maser (dotted line) and SIMS measurements (crosses) for diffusion of antimony after high-temperature implantation of boron into a wafer homogeneously doped with antimony. Dashed-dotted line: Approximation to the normalized vacancy profile [22]

lead to much more complex coupled systems of e.g. four partial differential equations, one ordinary differential equation, and two reaction equations [18] to describe the diffusion of one dopant species only. Such approaches, although needed to describe all effects occuring in dopant diffusion, are far too complex to be use in two- or even three- dimensional process simulation programs. However, basic experiments and theoretical investigations are still necessary to discriminate between different basic mechanisms involved. E.g. it is not yet finally decided whether in case of a homogeneous dopant profile, a vacancy gradient would lead to a diffusion of the dopants up or down the gradient: The first situation would result from a mechanism proposed by Maser [19] where the dopant and the vacancy simply interchange their places, whereas the second one would occur in case of the dopant atom and the vacancy forming a mobily pair [20]. This latter case, the so-called pair diffusion model, is presently preferred by most authors. Both approaches may result from atomistic investigations which only differ in the range of the attractive potential used [21]. Recent experimental investigations carried out at Erlangen are, however, in favour of the pair diffusion model: An inhomogeneous vacancy profile was generated by high temperature implantation of boron into a wafer which was homogeneously doped with antimony. Due to the elevated temperature and the vacancy gradient generated, the antimony diffuses during the implantation step. Fig. 2 shows the comparison between both approaches mentioned and SIMS measurements of the dopant profile. Only the pair diffusion model is capable to reproduce the shape of the measured profile, whereas the simple exchange of places between dopant atoms and vacancies does not reproduce the dopant pile-up observed near the surface.

Another important problem is the diffusion of dopant atoms across material interfaces, especially if these interfaces are moving like e.g. the interface between silicon and oxide in an oxidation process. E.g. in case of boron the segregation of the dopant at the interface leads to a large dopant flow from the silicon into the oxide across the moving interface. Assuming a stationary segregation condition with a segregation coefficient m, this flow across an interface which is moving with velocity v_{ox} is described by the Cauchy boundary condition

$$\begin{aligned} C_{ox} &= m \cdot C_{si} \\ D_{ox} \cdot \frac{\partial C_{ox}}{\partial n} &= D_{si} \cdot \frac{\partial C_{si}}{\partial n} + C_{si} \cdot v_{ox} \quad . \end{aligned} \tag{2}$$

Here, C_{ox} and C_{si} are the dopant concentrations at the oxide and at the silicon side of the interface, respectively. D_{ox} and D_{Si} are the diffusivities in the oxide and in the silicon, and $\partial/\partial n$ is the derivative normal to the interface.

An essential requirement for each process simulation program is that these kinds of Cauchy boundary conditions must be discretized in a way which allows for a sufficient control of the discretization errors of the fluxes across such moving interfaces. Otherwise the amount of dopant in silicon may be considerably wrong which would drastically influence critical electrical parameters such as the threshold voltage of the device. Again, this requirement must be met by the adaptive mesh which is anyhow necessary because of the changes in the dopant distribution. This problem is also addressed by the current activities mentioned in the preceeding section and will be published elsewhere [16].

The system of partial differential equations describing dopant diffusion is stiff and needs, therefore, to be solved implicitly. In consequence, after discretization of the relevant equations a large system of nonlinear equations results. In 3D simulations the dimension of the system may be up to 10^5. Especially if the number of nodes was minimized by use of Finite Boxes [23] or irregular refinements [24] during mesh generation the matrix of this system does not have an unperturbed band structure. This is a similar problem as in 2D and 3D device simulation. The development of suitable solvers for such systems is a topic of research in applied mathematics. In total, the solution of the diffusion problem in time-dependent 3D geometries is a challenge in terms of numerical algorithms, their parallelization, and the use of powerful computer equipment.

For the simulation of oxidation it is in addition to the effects mentioned above also necessary to calculate the diffusion of the oxidizing species through the oxide, their reaction with the silicon to form new oxide, and the mechanical problem resulting from the pressure

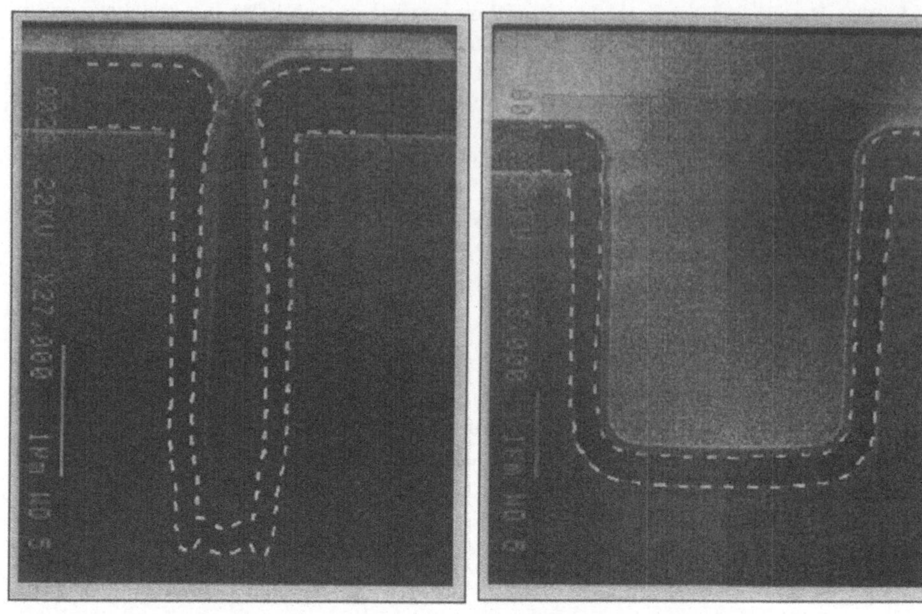

a) b)

Fig. 3: Comparison between experiment and simulation for the LPCVD deposition of oxide at a temperature of 570^0C, pressure of 0.1 Torr, gas flows of 60 sccm SiH_2 and 100 sccm O_2.

caused by masking layers and from the volume expansion of the newly generated oxide compared with the silicon which was converted into oxide. In order to obtain a sufficient accuracy of the simulations also in case of current complicated isolation structures like poly-puffered LOCOS, it is necessary to use a visco-elastic approach for the calculation of the mechanical problem. A considerable complication is that for the mechanical parameters, mainly the viscosity, and also for the diffusion coefficient of the oxidizing species and for their interface reaction rate, a stress dependency must be included.

LAYER DEPOSITION

Step coverage and trench filling are important issues of deposition processes used in silicon technology. Therefore, it is necessary to simulate such steps in two and in three dimensions. For low pressure chemical vapor deposition steps with pressures of about 0.1 Torr, the mean free path of a molecule in the reactive gas is far larger than the device feature sizes. Therefore, it is possible to model such processes assuming that the particle move along straight lines near the wafer surface. Far from the wafer, an isotropic distribution of the particle velocities is assumed due to thermal equilibrium. Furthermore it is assumed that a reactive molecule impinging on the wafer surface is absorbed with

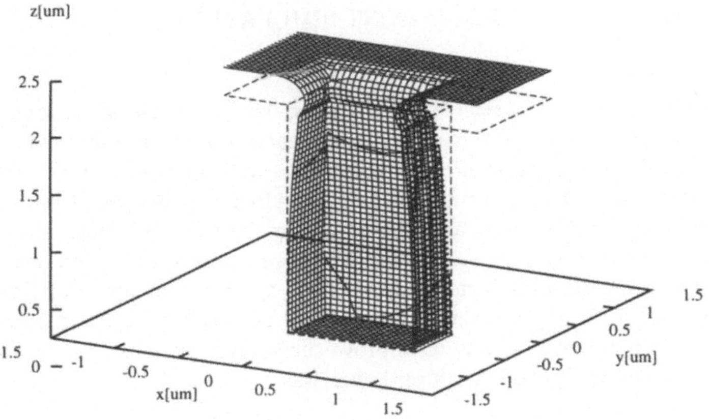

Fig. 4: 3D simulation of LPCVD deposition

a probability s and desorbed with a probability 1-s. s is called the sticking coefficient. The desorbed particles are emitted again from the surface according to a cosine angular distribution and either fly back into the free gas volume or hit the nonplanar wafer surface again at another position. In summary, the concentration of the reactive species at the wafer surface is result of a dynamical equilibrium between the incoming flux of particles arriving from the free gas volume or from other parts of the surface and the flux of desorbed particles. After a suitable discretization of the wafer surface, this concentration can therefore be calculated by solving a large system of linear equations [25]. The local growth rates are proportional to these concentrations and the wafer surface can then be easily updated in two dimensions using a modification [26] of the well-known string algorithm. More detailed experimental observations have shown that in case of deposition of low-temperature oxide, two reactive species contribute. Therefore, the model needs two sticking coefficients and the relative contribution of the two species as parameters [27]. In Fig. 3, comparisons between 2D simulations using this model and measurements are shown for LPCVD of SiO_2 at the same process conditions into trenches of two different widths. Very good agreement between experiment and simulation has been obtained.

Deep-submicron devices require the 3D simulation of such process steps. Whereas the physical model remains unchanged, the algorithmic problem is considerably more difficult in 3D. Main problems are the efficient calculations of view angles of each pair of surface planes and the 3D update of the topography. The considerable increase in size of the equation system to be solved causes less problems. In a current activity at Erlangen, a modified 3D string algorithm is used which shifts triangular planes according to the solution of the equation system mentioned above. In Fig. 4, an example for the 3D simulation of an oxide deposition into a trench is shown. Due to the plot program employed, the triangles used in the 3D string algorithm are not displayed in the figure.

EQUIPMENT SIMULATION

The homogeneity of process results across the wafer and between different wafers is an extremely important aspect which strongly influences the yield in semiconductor production. This aspect is getting even more important with increasing wafer sizes and more critical process specifications occuring with deep submicron devices. Therefore, it is desirable to obtain improvements by adjustment of the process specifications, optimization of process parameters or, even more efficient, by appropriate construction of the equipment. Again, an experimental approach would be very expensive in terms of time and money spent. Therefore, it is very attractive to use as far as possible modeling and simulation of the fabrication equipment already to improve the equipment in its design phase and later in order to use the equipment in an optimum way.

Presently, equipment simulation is widely used for the study of furnace processes, especially for the simulation of chemical vapor deposition (CVD), and for lithography and etching steps. The simulation of CVD reactors is based on general CFD codes like PHOENICS [28] or software specially developed for this purpose [29]. In addition to the standard equations for laminar flow also heat transfer due to radiation or convection and, most important, the chemical reactions occuring in the gas phase and at the surfaces must be considered.

In the following, results of simulations of metalorganic chemical vapor deposition (MOCVD) of the growth of $Al_xGa_{1-x}As/GaAs$ heterostructures in a horizontal reactor (AIXTRON) with rotating succeptor are outlined. In this process, metalorganic precursors of group III elements are introduced together with hydrides of group V elements. The carrier gas is H_2, with a gas pressure between 0.1 and 1 Bar. The model describes the flow of a chemically reacting mixture of gases by the Navier-Stokes equation. Complex models are being employed for the heat transfer between the gas, the solid walls of the reactor, and the susceptor, and for the radiative heat transfer. Furthermore, the transport of group III and group V species is included as well as the homogeneous and heterogeneous chemical reactions, the diffusion of the reaction products towards the growing layer, and the parasitic deposition of group III and group V elements on the heated parts of the reactor. More details of the model are given elsewhere [29,30]. Good agreement was obtained between the model developed and experimental data.

In Fig. 5, the influence of the wall temperature on the spatial inhomogeneity of the growth rate of $Al_xGa_{1-x}As$ layers along the flow direction and in perpendicular direction is shown. In the cold wall reactor with wall temperature about 300K the lateral homogeneity of the growth rate is good. Without wall cooling, however, the sidewall temperature increases to about 800 K, and the deposition rate drops considerably towards the wall due to considerable deposition of polycrystalline AlGaAs layers on the hot parts of the reactor wall and depletion of the gas mixture.

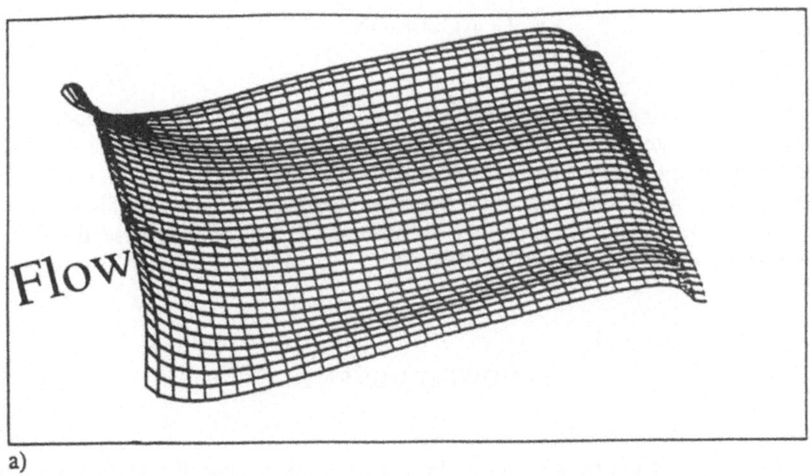

a)

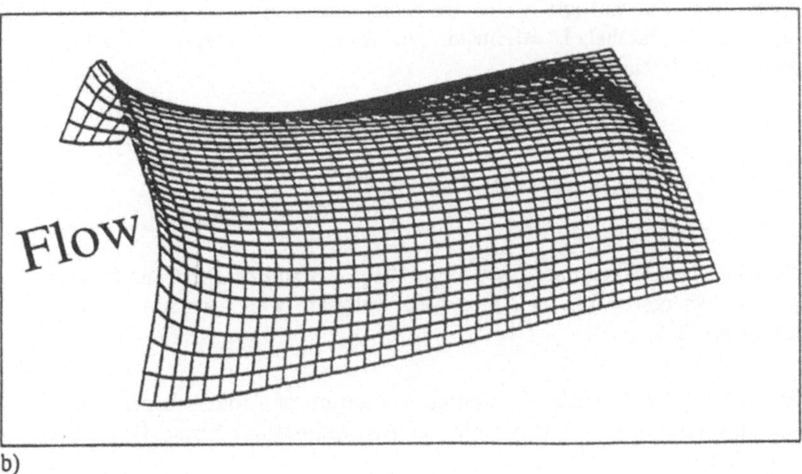

b)

Fig. 5: Spatial distribution of growth rate for $Al_xGaAl_{1-x}As$ layers: Upper figure for cold wall reactor (walls at room temperature), lower figure for reactor without wall cooling (T_W = 800 K). Susceptor temperature in both cases 1023K.

CONCLUSIONS

In this paper, a overview of activities at Erlangen in the field of process modeling and simulation is given. Some recent results have been outlined. The complexity of physical models for the description of current semiconductor fabrication steps and the need for two- or even three-dimensional simulation puts great demands on the numerical algorithms and computer systems used. Activities on high-performance computing are needed to provide the industry with appropriate tools in this field.

ACKNOWLEDGEMENTS

This work has been funded in part by the Bayerische Forschungsstiftung. Other parts of the work were carried out within ADEQUAT (JESSI project BT1B) and funded as ESPRIT project 7236. The authors would like to acknowledge the important contributions made by E. Bär, A. Barthel, L. Kadinski, Yu. Makarov, F. Meyer, P. Pichler, Martin Schäfer, and Michael Schäfer.

REFERENCES

[1] J.P. Biersack, L.G. Haggmark, A Monte-Carlo Computer Program for the Transport of Energetic Ions in Amorphous Targets, Nucl. Instrum. Meth. 174, (1980) pp. 257.

[2] M.T. Robinson, I.M. Torrens, Computer Simulation of Atomic-displacement Cascades in Solids in the Binary-Collision Approximation, Phys. Rev. B9, (1974) pp. 5008.

[3] I.R. Chakarov, R.P. Webb, Computer Simulation of B, P and As Channeling Profiles in Silicon, in: Proc. COSIRES 1992 (ed. J.P. Biersack), Berlin (1993).

[4] H. Stippel, S. Selberherr, Three Dimensional Monte-Carlo Simulation of Ion Implantation with Octree Based Point Location, in: Proceedings of the 1993 International Workshop on VLSI Process and Device Modeling (1993 VPAD), Nara, Japan, (1993) pp. 122.

[5] K.B. Winterbon, Ion Implantation in Inhomogeneous Materials, Appl. Phys. Lett. 31 10, (1977) pp. 649.

[6] J. Lorenz, W. Krüger, A. Barthel, Simulation of the lateral spread of implanted ions: theory, in: Proc. NASECODE VI (ed. J.J.H. Miller), Boole Press, Dublin, (1989) pp. 513.

[7] W.K. Hofker, D.P. Oosthoek, N.J. Koeman. H.A.M. De Grefte, Concentration Profiles for Boron Implantations in Amorphous and Polycrystalline Silicon, Radiation Effects 24, (1975) pp. 223.

[8] F. Jahnel, H. Ryssel, G. Prinke, K. Hoffmann, K. Müller, J.P. Biersack, R. Henkelmann, Nucl. Instrum. Meth. 182/183, (1981) pp. 223.

[9] M. Kuhn, Analytische Modelle für Implantationsverteilungen in Silicium, Diplomarbeit, Lehrstuhl für Elektronische Bauelemente, Universität Erlangen-Nürnberg, (1987).

[10] A.F. Tasch, H. Shin, C. Park, An Improved Approach to Accurately Model Shallow B and BF_2 Implants in Silicon, J. Electrochem. Soc. 136 (3), (1989) pp. 810.

[11] H. Ryssel, J. Lorenz, K. Hoffmann, Models for the Implantation into Multilayer Targets, Appl. Phys. A 41, (1986) pp. 201.

[12] R.J. Wierzbicki, J.P. Biersack, A. Barthel, J. Lorenz, H. Ryssel, Reflection Approach for the Analytical Description of Light Ions Implanted into Bilayer Structures, in: Proc. COSIRES 1992 (ed. J.P. Biersack), Berlin (1993).

[13] J. Lorenz, C. Hill, H. Jaouen, C. Lombardi, C. Lyden, K. de Meyer, J. Pelka, A. Poncet, M. Rudan, S. Solmi, The STORM Technology CAD System, in: Technology CAD Systems (eds. F. Fasching, S. Halama, S. Selberherr), Springer Verlag, Wien, New York, (1993) pp. 163.

[14] R.J. Wierzbicki, J. Lorenz, H. Ryssel, Advanced Analytical Models for the Multidimensional Simulation of Ion Implantation, to be published.

[15] J. Lorenz, R.J. Wierzbicki, Efficient Multidimensional Simulation of Ion Implantation into Multilayer Structures, in: Proc. of the 1993 International Workshop on VLSI Process and Device Modeling (VPAD 1993),Nara, Japan, (1993) pp. 84.

[16] J. Lorenz, F. Meyer, to be published.

[17] M. Orlowski, Progress in Process Simulation for Submicron MOSFETs, in: Simulation of Semiconductor Devices and Processes Vol. 3 (eds. G. Baccarani, M. Rudan), Tecnoprint, Bologna, (1989) pp. 393.

[18] R. Dürr, Beschreibung der Dualen Diffusion von Dotieratomen in Silicium für die Anwendung in der Prozeßsimulation, Dissertationsschrift, Universität Erlangen-Nürnberg (1991).

[19] K. Maser, Dotandendiffusion und -drift im Silizium mit inhomogen verteilten Vakanzen, Experimentelle Technik der Physik 34, (1986) pp. 213.

[20] M. Orlowski, Unified Model for Impurity Diffusion in Silicon, Appl. Phys. Lett. 53, (1988) 1323.

[21] S. List, P. Pichler, H. Ryssel, Atomistic Evaluation of Diffusion Theories for the Diffusion of Dopants in Vacancy Gradients, in: Simulation of Semiconductor Devices and Processes Vol. 5 (eds. S. Selberherr, H. Stippel, E. Strasser), Springer Verlag, Wien, New York, (1993) pp.97.

[22] P. Pichler, Direct Experimental Evidence for Diffusion of Dopants via Pairs with Intrinsic Point Defects, Appl. Phys. Lett. 60 (8), (1992) 953.

[23] A.F. Franz, G.A. Franz, S. Selberherr, C. Ringhofer, P. Markovich, Finite Boxes - A Generalization of the Finite Difference Method Suitable for Semiconductor Device Simulation, IEEE Trans. El. Dev. ED-30 (9), (1982) pp. 1070.

[24] P. Conti, Grid Generation for Three-dimensional Semiconductor Device Simulation, in: Series in Microelectronics Vol. 12 (eds . W. Fichtner, W. Guggenbühl, H. Melchior, G.S. Moschytz), Hartung-Gorre-Verlag, Konstanz, (1991).

[25] H. Wille, E. Burte, H. Ryssel, Simulation of the Step Coverage for Chemical Vapor Deposited Silicon Dioxide, J. Appl. Phys. 71, (1992) pp. 3532.

[26] J. Lorenz, C. Hill, H. Jaouen, C. Lombardi, C. Lyden, K. de Meyer, J. Pelka, A. Poncet, M. Rudan, S. Solmi, The STORM Technology CAD System, in: Technology CAD Systems (eds. F. Fasching, S. Halama, S. Selberherr), Springer Verlag, Wien (1993) pp. 163.

[27] H. Wille. E.P. Burte, A Dual Sticking Coefficient Chemical Vapor Deposition Model, in: Proc. ESSDERC '92 (eds. H.E. Maes, R.P. Mertens, R.J. Van Overstraeten), Elsevier, Amsterdam, (1992) pp. 503.

[28] W.H. Mills, High Temperature Furnace Simulation with Surface to Surface Radiation, PHOENICS Journal of Computational Fluid Dynamics 4 (4), (1991) pp. 389.

[29] F. Durst, L. Kadinski, M. Peric, M. Schäfer, Numerical Study of Transport Phenomena in MOCVD Reactors Using a Finite Volume Multigrid Solver, J. Cryst. Growth 125, (1992) pp. 612.

[30] Yu.N. Makarov, E.V. Subashieva, A.N. Zabolotskikh, A.I. Zhmakin, Numerical Simulation of Flow and Heat Transfer in Epitaxial reactors, in: Numerical Methods in Thermal Problems VIII (ed. R.W. Lewis), Pineridge Press, (1993) pp. 1298.

NUMERICAL SIMULATION OF CRYSTAL GROWTH PROCESSES

G. Müller
Univ. Erlangen-Nürnberg, Institut für Werkstoffwissenschaften,
Martensstraße 7, D 91058 Erlangen, Germany

SUMMARY

The present status of numerical simulation of crystal growth processes, especially by the Czochralski technique, is discussed by comparison of numerical and experimental results. It is shown in which cases the convective heat transport in the melt has to be considered in order to give realistic numerical results. Furthermore, examples were presented in which 2-dimensional simulations agree with experiments and cases where a 3-dimensional simulation is necessary. Finally, it will be shown that for a simulation of doping distribution in the crystal, i.e. segregation, the crystal melt interface has to be treated as a "moving boundary".

CRYSTAL GROWTH AS KEY TECHNOLOGY TO MICROELECTRONICS

In the world economy, the electronics industry is the largest and by far the fastest growing manufacturing industry. The worldwide investment is currently estimated to be in the range of $900 billion, much larger than automotive or steel. The semiconductor portion which forms the basis of elctronics industry is in the $60 - 70 billion range (source: C.R. Barret/Intel Corp. [1]). The key material to electronics is the single crystalline silicon. These crystals are produced with a volume of about 5000 metric tons a year and a market value of more than one billion $. The silicon wafers are cut from the single crystals produced from the melt, using special techniques called "crystal growth" at temperatures slightly over 1420°C by defined control of the liquid-solid phase transition. The most important technique is the pulling of a crystal from a crucible (Czochralski method).

The size of such a crystal, more than one meter in length as depicted in Fig.1, demonstrates that in the Czochralski process melt volumes weighing more than 50kg have to be handled and controlled. The demand for larger crystal dimensions arises from the desire to increase the profitability in manufacturing the crystals, and the yield of electronic devices fabricated on the crystal wafers.

Fig.2 shows the historical progression and the future trend of wafers, in crystal diameter and complexity of electronic devices (here DRAMs, Dynamic Random Access Memory). The tremendous increase of the number of transistors, e.g. DRAMs per chip can only be achieved by a decreasing of the line-width of the device structures from presently 0.4μm to 0.2 or even 0.15μm in the year 2000 [2]. These goals for the devices are setting the landmarks for the development of crystal growth. It is foreseeable that the yield of device fabrication would become uneconomically small with increasing integration, if the crystal quality (e.g. the density of defects) remains the same. Therefore, the defect density of crystals and non-uniformities must be reduced simultaneously with the increase of the crystal dimensions. Furthermore, the tolerances of specifications for the electronic properties such as resistivity in the micro-scale (lateral and longitudinal) must also be narrower than the current standard.

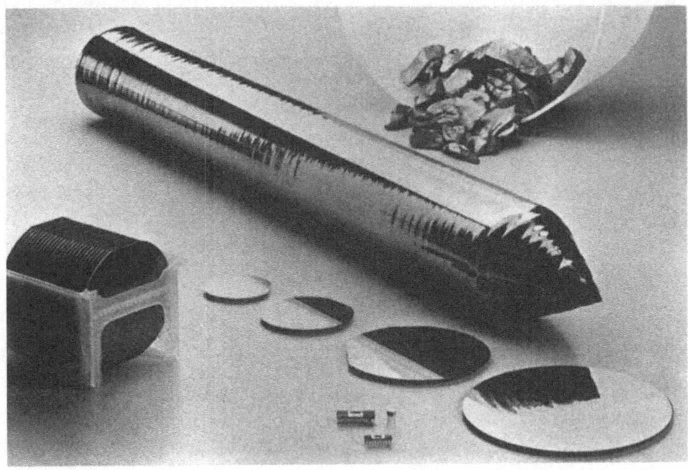

Fig.1 Silicon single crystal (ϕ = 154mm) grown by the Czochralski technique from a Silica crucible containing the poly silicon (in the back). Silicon wafers with various diameters and integrated circuits (front).
Source: Werkphoto Wacker Chemitronnic.

Altogether, this is a rather tough task for scientists and engineers working in crystal growth, if one considers the complexity of the Czochralski process (see next section). The task involves an upscaling of existing equipment as well as R + D work on new processing techniques (like continous melt feed). To achieve this goal a large number of parameter variations is necessary, which is very time consuming and expensive.

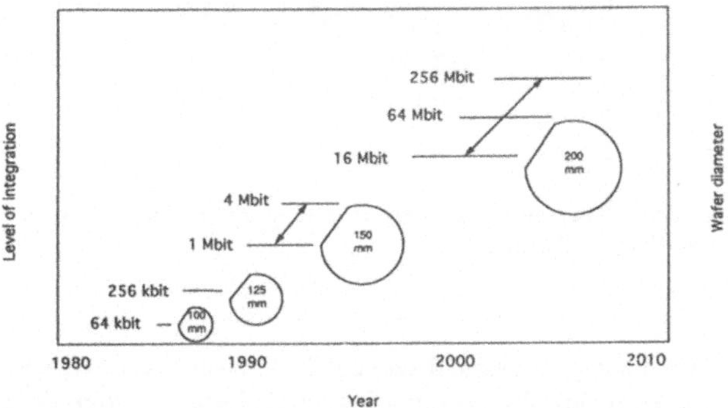

Fig.2 Level of integration in terms of DRAMs per chip and diameter of Si crystal or wafer, respectively, versus time scale [3].

In this situation, numerical simulations can effectively support the developments for improved crystal growth processes, because it allows, in principle, an easy way to change material properties, equipment geometry and growth parameters. However, the results of numerical simulations can be used for industrial purposes only, if the model is close to the realistic production process.

The aim of this paper is to present the state of art of our modelling within the FORTWHIR project and in comparison to the international literature. For such purpose we have first to consider in detail the features of the Czochralski crystal pulling technique and the mathematical description of the occuring physical processes.

CRYSTAL PULLING FROM THE MELT: CZOCHRALSKI METHOD

The principle of crystal pulling from the melt dates back to the early work of Czochralski in 1917. But it lasted until the fifties, when the first semiconductor crystals of Ge and later Si were grown by this principle. Later, with increasing importance of optoelectronic devices

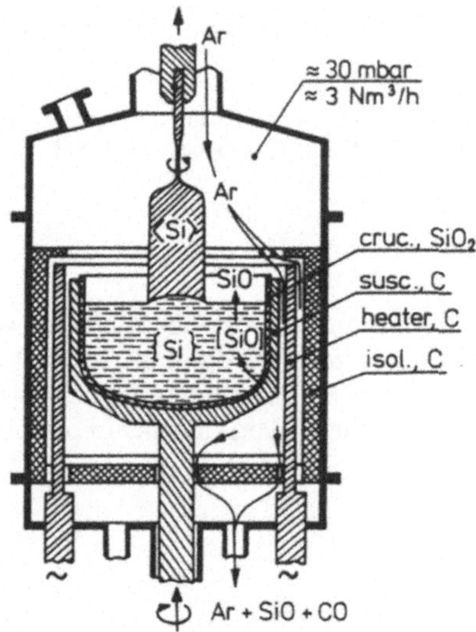

Fig.3 Principle of the set-up of an industrial Czochralski pulling apparatus for the growth of Si-crystals. Source: Zulehner, Wacker Chemitronic.

(light emitting and laser diodes), compound semiconductors such as GaAs and InP were also grown industrially by the Czochralski method using a liquid encapsulation of the melt, called LEC process. Today more than 80% of all semiconductor crystals worldwide are grown by the Czochralski method. The principle of the set up of a modern Czochralski pulling apparatus for

Si crystal growth is depicted in Fig.3. The silicon melt is heated by a graphite resistance heater in a silica (SiO_2) crucible. The crucible is kept in a stabilizing graphite susceptor rotating axially to provide axisymmetric heating conditions. A silicon single crystal seed of a certain crystallographic orientation is dipped into the melt and pulled out under rotation (counter to the crucible rotation). After the growth of a conic portion, the crystal diameter is controlled to maintain constant given by the wafer diameter. During growth, the processing chamber (water cooled-steel walls) is flushed by a continous Ar-stream, which is used to cool the crystal and to remove the gaseous SiO evaporating from the free Si-surface.
The transport of SiO and its incorporation into the Si-crystal are playing a key role in the Czochralski process. It should be considered, therefore, in more detail. The SiO_2-crucible is eroded by the Si-melt which gives a continuous contamination of the Si-melt with oxygen (presumeably as SiO). Oxigen is transported in the melt by convective processes and most of it evaporates through the free surface. But a certain amount of some 10^{17} atoms/cm^3 is transported to the growth interface and incorporated into the Si-crystal. This oxygen doping of the crystal is an important prerequisite for the semiconductor processing technology because of two effects. Firstly, oxygen precipitates during cooling by forming crystallographic defects (dislocation loops and stacking faults) which act as gettering centers for residual metallic impurities in the silicon wafers during device processing. Secondly, oxygen hardens the silicon lattice which gives the thin wafers (about 300µm) the necessary mechanical strength during device processing, especially at high temperatures (~1000°C). For these reasons, the oxigen concentration and uniformity in the crystal must be within very narrow tolerances. This can only be achieved by a quantitative control of the transport processes and other growth parameters (for further details see the review article of Zulehner and Huber [2]).
Numerical modelling of the Czochralski process is strongly motivated by the need in the control of the oxygen transport and incorporation .

NUMERICAL MODELLING OF THE CZOCHRALSKI PROCESS

The general task of process modelling in crystal growth can be defined as follows: Correlation between crystal properties and all relevant growth parameters. In the case of Czochralski growth of Si, this means correlation between the

Relevant crystal properties:

* oxygen concentration
* doping concentration
* microdefects
* distribution of oxygen, dopants and microdefects,

Growth parameters:

* temperature profile in melt and crystal
* rotation rates of crystal and crucible
* pulling rate
* shape of crystal-melt interface
* position of crystal-melt interface
* shape of melt meniscus close to the growth interface
* geometry of crucible, i.e. melt
* external fields, i.e. magnetic fields

The correlation between the crystal properties and the growth parameters are given by the governing physical mechanisms which are described mathematically by the corresponding differential equations:

* Heat transport by conduction, convection (in all fluid phases) and radiation within the whole equipment. The convection can be free (thermal and / or solutal buoyancy) or forced by rotation.
* Mass transport (dopants, oxygen, impurities) in the melt or gas environment.
* Free boundary problem (Stefan problem) at the interface of melt and solid crystal, segregation effects at the phase boundary and incorporation of dopants in the crystal.
* Free surface of fluids which lead to static surface tension effects (shape of menisci) and dynamic effects like surface driven convection (Marangoni convection) which can have thermal or solutal origin.
* Defect formation (generated e.g. by mechanical strain) and diffusion in the solid crystal.

Furthermore, it should be kept in mind that all material parameters are not constant, at least they show a dependence on temperature.
The explicit mathematical formulation can be found elsewhere in detail [4-6]. It consists of a set of coupled non-linear differential equations with certain boundary conditions, part of it is a free-boundary (solid-liquid interface of the growing crystal). Models which include the whole

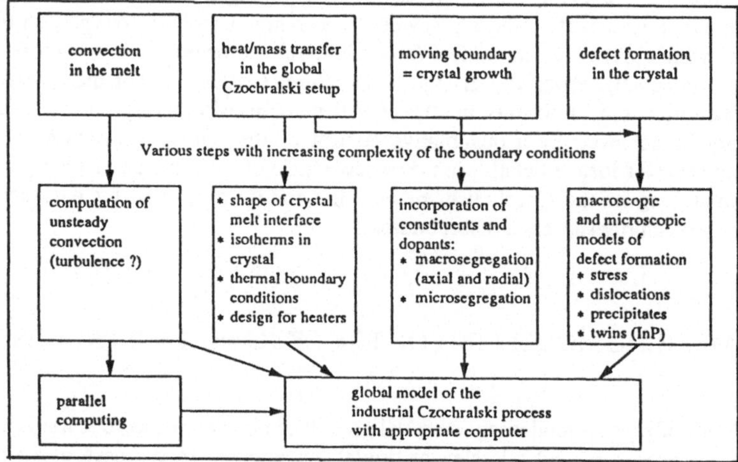

Fig.4 Strategies for computation of the complex Czochralski process.

processing equipment as sketched in Fig.3 are called global models. Others consider only certain parts, e.g. melt and growing crystal. The set of differential equations is solved by discretization using various numerical methods like finite element, finite differences or finite volume. Several projects within FORTWHIR are working on the development of such codes (see special contributions in this book).Worldwide only a few groups [7-9] are working in the field of modelling the global Czochralski process because of the complexity of the problem needing very high computer capacity and CPU time. This is illustrated by several examples, which will be given in the next section. It will be shown, that one key point for successful numerical simulations is a correct description of the heat and mass transport within the melt which are described by the time dependent equations for conservation of momentum, energy and mass. This is also the most time consuming (concerning computer recources) part of the simulation because convection, which plays an important role in the transport mechanisms, is strongly time dependent or even turbulent in industrial crystal growth situations. The numerical algorithms which are used up to now for this purpose are not efficient enough to simulate the transport mechanisms, especially convection in industrial crystal growth, even on modern supercomputers. Therefore, the computation time for simulations has to be decreased considerably. One way to achieve this aim is to employ more advanced computer architectures e.g. parallel computers. At the same time it is also of great importance to improve old and develop new numerical algorithms for a full exploitation of the capabilities of these new computer architectures. These are important tasks which were studied by several groups within projects of FORTWHIR. The strategies for computation of the complex industrial Czochralski process are summarized in Fig.4.

In order to develop reliable models which can capture all essential physical phenomena of the crystal growth processes, comparison of numerical results to experimental data is indispensable. For that reason, experiments have to be performed, both in the field of real crystal growth (e.g. temperature measurements within an industrial silicon Czochralski furnace) as well as experimental modelling (e.g. investigation of model fluid convection within a heated cavity).

The next section gives several examples of such a procedure.

RESULTS OF NUMERICAL SIMULATIONS AND COMPARISON TO EXPERIMENTS

In this section we will demonstrate with three examples of numerical simulations the state of the art of our models by comparing the numerical results with experiments. With these examples it will also be shown, that certain simplifications of the models can be made, if one is only interested in special features and neglect the details of the process.

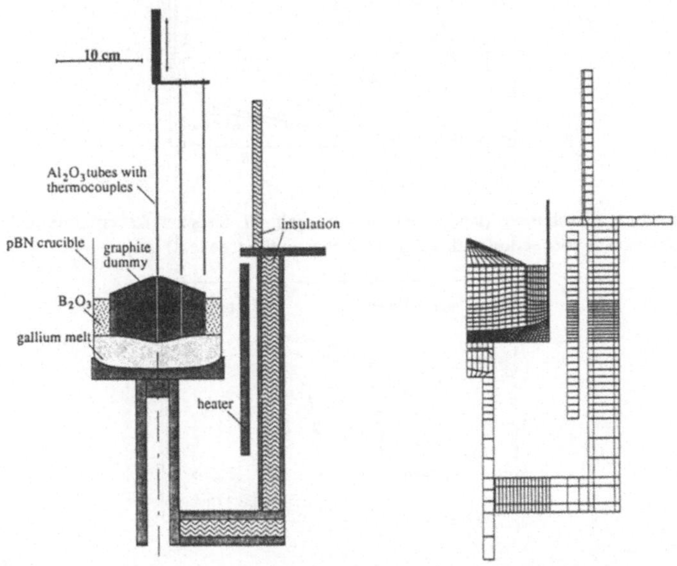

Fig.5 Sketch of a model experiment for temperature measurements in an industrial Czochralski (LEC) apparatus (left) and numerical grid (2-dimensional) of this set up (right) [10].

(i) Convection in the melt

The mathematical treatment of convection in the melt, especially by solving the non-linear Navier-Stokes equations causes big numerical problems. It is, therefore of interest to study the question, whether convection in the melt has to be considered or not. We have done this by using two types of numerical models: one which includes convection (II) and one without convection (I). Model I is a global conduction-radiation model (without convection) based on a commercial finite element program ABAQUS, assuming axi-symmetry and quasi-steadiness. The numerical grid is shown in Fig.5. It contains all important apparatus details of

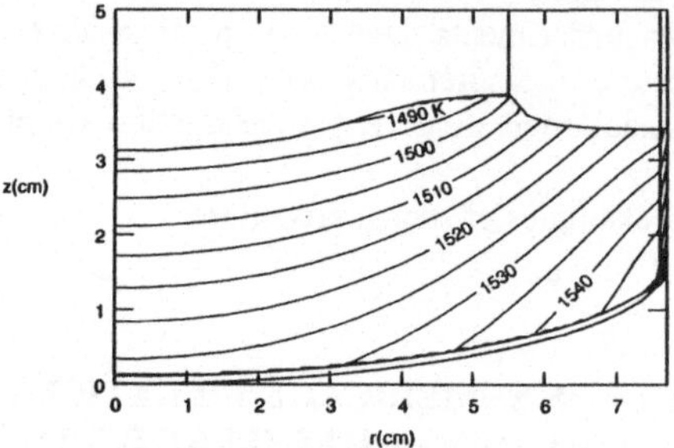

Fig.6 Temperature distribution (isotherms) in the melt of the Czochralski crucible calculated with a conduction-radiation model (ABAQUS) without convection [10].

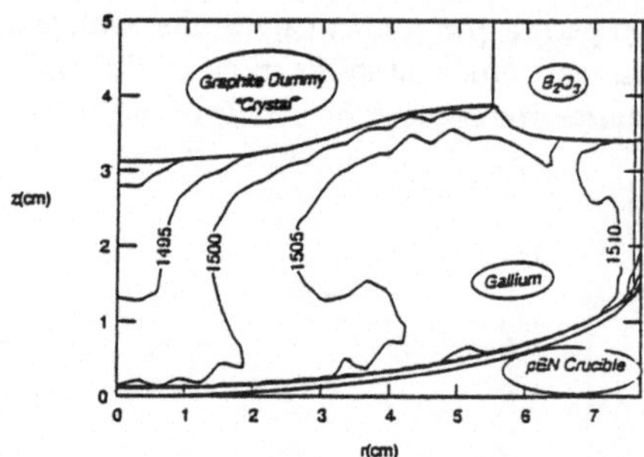

Fig.7 Temperature distribution (isotherms) in the melt of Czochralski crucible calculated under consideration of convection with a 2-dimensional model (see text) [10].

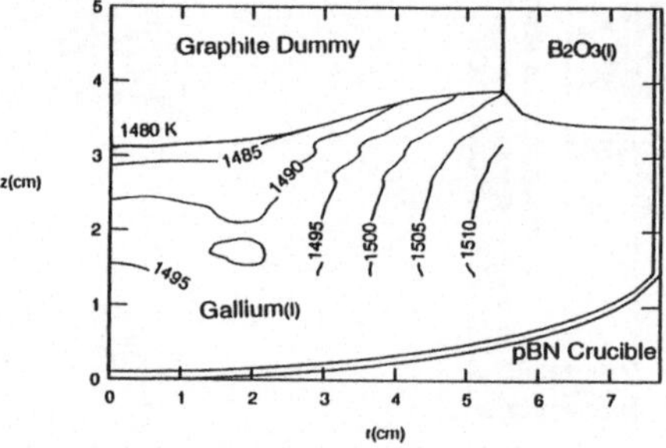

Fig.8 Measured temperature distribution (isotherms) in the melt of the Czochralski crucible (by the set up shown in Fig.5).

the industrial Czochralski configuration (LEC) which is shown on the left hand side of the figure. The experimental data (temperature distributions in melt and crystal) were obtained by thermocouple measurement. We were using a graphite dummy with axial holes for the thermocouples to simulate the GaAs crystal. The GaAs melt was replaced by Ga. Both changes do not considerably alter the thermal conditions of the industrial LEC-process. The calculation was carried out by iterating the power input to the graphite heater so that the solid / melt / encapsulant trijunction is at a preset temperature. A typical numerical result of the temperature distribution (isotherms) is shown in Fig.6. Such a calculation takes about 15 minutes of CPU time on a HP Apollo 720 workstation. For more details, see Koai et al. [10]. Convection was considered in model II by calculating the heat transfer in the melt separately with another commercial code, called STAR-CD. This is a finite volume code based on the SIMPLE algorithm. The heat transfer boundary conditions for the melt were obtained from the results of the global model I by matching the temperatures at the boundaries of the melt. This way we combined both models to obtain a global convective-radiative model. For more details see Baumgartl et al. [11]. About 5 to 10 iterations between the two codes are necessary before a globally converged solution was available. This takes about 10 hours of CPU time on a HP Apollo 720 workstation. A typical result of the temperature distribution in the melt is shown in Fig.7. Both results, obtained without considering convection (model I, Fig.6) and with convection (combined model I/II, Fig.7) are compared with the experimentally determined isotherms in the melt (Fig.8). The comparison clearly shows the following: The conductive-radiative model (I) cannot predict the temperature distribution in the melt closely to reality. However, it follows from a more detailed study [10], that it is well suited to predict temperature fields and heat transfer in the solid parts of the Czochralski apparatus. The combined convection model (I/II) gives a prediction of the temperature distribution in the melt which agrees fairly good with the measured one.

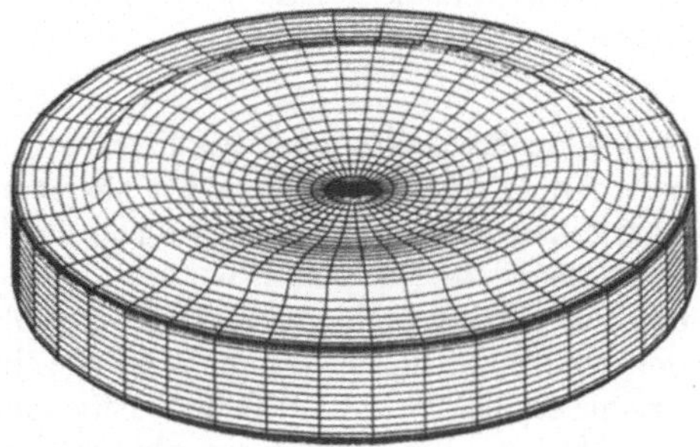

Fig.9 Numerical grid for a 3-dimensional calculation of convective transport in the melt of the Czochralski crucible [10].

So far we have only used axy-symmetric calculations. The next question to be answered is whether there are cases of interest in which the calculation has to be carried out 3-dimensionally.

(ii) 3-dimensional calculations

First we consider again the Czochralski configuration treated in the previous section (i). Now we are using a 3-dimensional Navier-Stokes solver based on the finite volume method and a multigrid technique with a fully vectorized structure [10, 14]. The 3-dimensional grid for the melt is shown in Fig.9. The computation takes 2-3 hours of CPU time on a CRAY-YMP supercomputer to establish the main features of flow pattern and temperature field. Again we have combined this model with the conductive-radiative global model I (ABAQUS) by matching the temperatures at the boundaries of the melt. After 5 to 6 coupling iterations we achieved a converged solution with differences in the boundary temperatures between the runs below 1K. A typical result of such a 3-dimensional simulation is shown in Fig.10. The temperature distribution is in good correlation to the experimental result (Fig.8). This

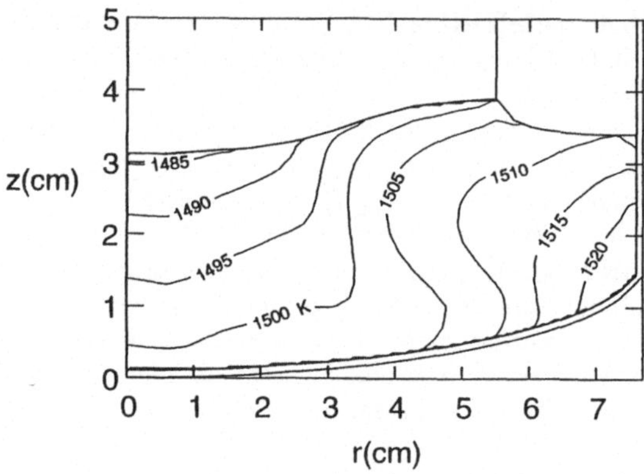

Fig.10 Temperature distribution (isotherms) in the melt of the Czochralski crucible calculated under consideration of convection with a 3-dimensional model (see text) [10].

computational scheme, although scientifically more accurate, requires more expensive computing resources at the present time. Since the difference in the predictions by this method and the previous 2-dimensional calculations is marginal (compare Fig.7), the 2-dimensional model might suffice for most industrial applications. However, it should be mentioned, that the 2-dimensional calculations are causing problems for high crucible rotation rates (e.g. 30rpm). In that case obviously 3-dimensional structures of the flow are causing the 2-dimensional solutions to become oscillatory. Two different solution branches were obtained (bifurcation) from which one had to be selected by physical arguments [11]. With 3-dimensional calculations this problems did not occur, even with high crucible rotation rates.
Next we consider the structure of flows and wave patterns in a silicon melt in an industrial Czochralski apparatus (Leybold EKZ 1300). For simplicity, we discuss here only the situation without the silicon crystal. The flow patterns were investigated experimentally by using a pair of thermocouples dipping into the melt at the same radius but separated by an angle. Different crucible rotation rates were applied. Analysis of the Fourier spectra of the temperature signals revealed flows with m-folded, precessing wave patterns. The integer m varies between 2 and 5 depending on the crucible rotation rate. For more details see Seidl et al. [12]. This complex behaviour of the flow which may disturb the growing crystal is speculated in the literature to be either a baroclinic instability, Küppers-Lortz instability or a Hopf bi-furcation. More

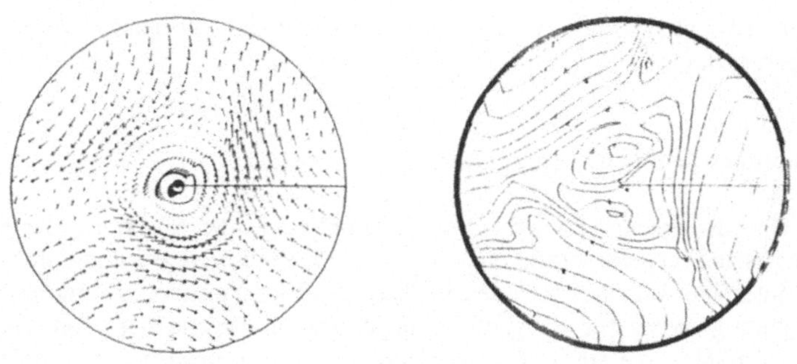

Fig.11 Calculated flow pattern (left) and temperature distribution (right) in a horizontal section of a Czochralski melt just below the crystal interface. The distance of isotherms is 3K [12].

insight into the mechanism was expected from quantitative numerical simulation results. The mathematical model must be 3-dimensional, of course [12]. The same numerical finite volume code was used as mentioned above and run on a CRAY-YMP. Even by using a

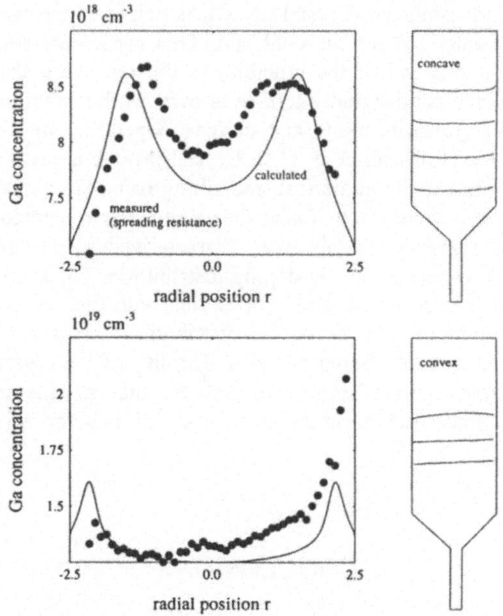

Fig.12 Calculated and measured doping (Ga) profile across a 2" Ge wafer grown by the vertical gradient freeze method with concave (above) and convex (below) interface shape [13].

relatively coarse grid, the same m-folded azimuthal waves could be calculated as experimentally observed. Fig.11 shows a typical example of a three-fold symmetry. From the results of these numerical simulations we get a lot of additional information about the flow compared to experimental data [12].

(iii) Moving boundary problem - segregation

In the global models discussed in (i) and (ii) we have only considered heat transfer problems. Transport of species and its incorporation into the crystal including segregation phenomena in the solid-liquid interface were not included in these models. Now we are interested in the distribution of doping atoms (or oxygen) in the crystal in dependence from the growth conditions. For simplicity, we study this effect not in the Czochralski but in a cylindrical vertical gradient freeze (Bridgman) configuration. This model contains all features of the doping incorporation of a Czochralski situation but has less complex physical conditions, e.g. low Rayleigh number, no rotation, no free melt surface and no melt meniscus. The model consists of a finite volume code which was developed in collaboration with the Institute of Fluid Mechanics at the University of Erlangen-Nürnberg. It solves the temperature field in melt, crystal and crucible as well as convection and dopant distribution in the melt axi-symmetrically. The model considers a cylindrical crucible, surrounding the melt and crystal. The temperature profile of the heater is taken from experimental data. The heat transfer to the crucible is described by heat transfer coefficients. The solid-liquid interface is fixed to the melting point-isotherm. Crystal growth is achieved by moving the temperature profile of the heaters axially toward the cold end. This causes the solid-liquid interface to move from the seed (bottom end of the crucible) to the top until the total melt has been solidified. The shape of the solid-liquid interface is part of the solution (Stefan problem). It varies between concave (towards melt) and convex depending on the thermal boundary conditions. For details see Hofmann et al. [13]. Crystal growth experiments were carried out in our laboratory in parallel to the numerical modelling using Ge doped with Ga as a model substance [13]. The crystals were grown under the same thermal boundary conditions as used for the calculations. The grown crystals were characterized by microscopic and electrical analysis, especially with respect to the Ga-doping distribution. The shapes of the interface, as well as the doping distributions are in good correlation with the numerical simulations [13]. Fig.12 gives two examples of the radial Ga distribution across a Ge-wafer. The results demonstrate that there is a much better radial uniformity of the crystal for certain thermal boundary conditions (here "convex" interface) than for others. This shows clearly that the model can be used to find the optimum growth conditions for achieving homogeneous crystals.

CONCLUSIONS

Numerical modelling of crystal growth processes can be a valuable tool for the development of improved industrial crystal growth processes like Czochralski. But numerical simulations cannot fully replace R + D experimentation in industrial crystal growth. The strength of numerical simulations consists in the possibility to carry out series of parameter variations much more efficiently than by experiments. Global modelling of all important features of the process is possible - in principle. However, it is more efficient to reduce the complexity to that level which is necessary to provide a realistic result in a certain case. The use of numerical modelling for industrial problems is still limited by the low efficiency of numerical codes and computer capacity. Promising developments as block structuring, parallelizing and the use of transputer computers are studied within the FORTWHIR research projects.

ACKNOWLEDGEMENTS

The author gratefully acknowledges the contributions of the following co-workers of our crystal growth laboratory: Dr. J. Baumgartl, Dr. A. Bune, Dr. D. Hofmann, Dr. T. Jung, Dr. K. Koai, G. McCord, Dipl.-Ing. A. Seidl and Dipl.-Phys. H.-J. Leister from the Fluid Mechanics Dpt. The work was supported by the Bavarian Ministery of Science and Culture, the Bavarian Research Foundation, the Volkswagen Foundation and the industrial partners Leybold AG, Freiberger Elektronikwerkstoffe and Wacker Chemitronic.

REFERENCES

[1] BARRET, C.R.: "Silicon Valley, What Next?", Plenary adress given at the 1993 MRS Spring Meeting, MRS Bulletin, Juli 1993, 3-10.

[2] ZULEHNER, W. and HUBER, D.: "Czochralski-Grown Silicon", Crystals Vol. 8, J. Grabmaier (ed.), Springer Verlag, Heidelberg (1982), 1-143.

[3] BULLIS, W.M. and O'MARA, W.C.: "Large-Diameter Silicon Wafer Trends", Solid State Technology, April 1993, 59-65.

[4] HURLE, D.T.J.: "Crystal Pulling from the Melt", Springer-Verlag, Heidelberg 1994.

[5] MÜLLER, G.: "Crystal Growth from the Melt", CRYSTALS Vol.12, H.C. Freyhardt (ed.), Springer Verlag, Heidelberg 1988.

[6] MÜLLER, G. (ed.): "Improvement of Manufactoring Processes by Semiconductor Crystala by Modelling of the Transport Processes", University Erlangen-Nürnberg, Final Report of a project sponsored by the Volkswagen Foundation (July 1993).

[7] KINNEY, T.A., BORNSIDE, D.E. and BROWN, R.A.: "Quantitative assessment of an integrated hydrodynamic thermal-capillary model for large diameter-Czochralski growth of silicon: comparison of predicted temperature field with experiment", J. Crystal Growth 122 (1993) 413-434.

[8] DUPRET, F., NICODEME, P., RYCKMANS,Y., WOUTERS and CROCHET, M.J.: "Global modelling of heat transfer in crystal growth furnaces", Int. J. Heat Mass Transfer 33 (1990) 1849-1871.

[9] MIYAHARA, S. et al.: "Global heat transfer model for Czochralski crystal growth based of difuse-gray radiation", J. Crystal Growth 99 (1990) 697-701.

[10] KOAI, K., SEIDL, A., LEISTER, H.-J., MÜLLER, G. and KÖHLER, A.: "Modelling of thermal fluid flow in the Czochralski process (LEC) and comparison with experiments", J. Crystal Growth 1993 (accepted).

[11] BAUMGARTL, J., BUNE, A., KOAI, K., MÜLLER, G. and SEIDL, A.: "Global simulation of heat transport, including melt convection in a Czochralski crystal growth process - combined finite element / finite volume approach", E-MRS 1993 Spring Meeting, Strasbourg (F) 4-7 May 1993, Symposium F.

[12] SEIDL, A., McCORD, G., MÜLLER, G. and LEISTER, H.J.: "Experimental observation and numerical simulation of wave patterns in a Czochralski silicon melt", J. Crystal Growth (1993) submitted.

[13] HOFMANN, D., JUNG, Th. and MÜLLER, G.: "Growth of 2 inch Ge:Ga crystals by the dynamical vertical gradient freeze process and its numerical modelling including transient segregation", J. Crystal Growth 128 (1993) 213-218.

[14] LEISTER, H.-J. and PERIC, M.: J. Crystal Growth 123 (1992) 567-574.

QUANTUM MONTE CARLO SIMULATIONS AND WEAK-COUPLING APPROXIMATIONS FOR THE THREE-BAND HUBBARD MODEL

R. Putz, H. Endres, A. Muramatsu, and W. Hanke
Physikalisches Institut der Universität Würzburg
Am Hubland, D-97074 Würzburg, Germany

SUMMARY

We study the three-band Hubbard model for the cuprate superconductors using Quantum Monte Carlo simulations and the fluctuation exchange approximation including a self-consistent many-body renormalization of the single-particle propagator. The energy dispersion of the low-lying one-particle excitations is in very good agreement with spectroscopic measurements for a hole doping concentration $\delta = 0.25$ of the CuO_2 planes. At the same doping, we examine the spin susceptibility yielding a non-ordered phase with short-range incommensurate fluctuations near the antiferromagnetic wave vector. In the superconducting regime, the interaction vertex for the pairing correlation function in the extended s-wave channel shows a maximum for $\delta = 0.20$, resembling the dependence of transition temperature T_c on doping.

INTRODUCTION

Strongly correlated electron systems, such as the high-temperature cuprate superconductors, have been studied in recent years with considerable effort using a wide variety of theoretical approaches [1]. Quantum Monte Carlo (QMC) simulations provide a direct non-perturbative calculation of correlation functions starting from microscopic models of such correlated systems [2, 3, 4, 5, 6]. On the other hand, this method is limited by present computing capabilities and more fundamental difficulties, commonly known as the fermionic sign problem [7, 8], to relatively small lattices and high temperatures. Therefore, we employ a more conventional perturbative expansion up to intermediate values of the electronic Coulomb interaction as a useful complement to QMC. In the present work, we apply the fluctuation exchange (FLEX) approximation [9, 10, 11] which symmetrically (in particle-hole and particle-particle channels) takes into account the interaction of electrons with density, magnetic, and superconducting pair-fluctuations for the single particle propagator. These fluctuations are expected to be particularly important in regions of competing phases, as for the cuprate superconductors. In order to perform this perturbation expansion in a well-controlled manner, the results must be compared with, in principle, exact QMC data in a common parameter regime of the microscopic model. In the present work, QMC simulations and perturbative weak-coupling approximations are employed for the three-band Hubbard model [12, 13] to calculate the one-particle excitation spectrum (i.e. the band-structure), the static magnetic susceptibility, and the superconducting interaction vertex. For the low-energy one-particle excitations, we are

in very good agreement with experimental measurements of the Fermi surface and Fermi velocity for both of the above mentioned theoretical techniques. In weak-coupling approximations, the self-consistent inclusion of many-body fluctuation effects is indispensable in order to avoid unphysical magnetic instabilities of the normal phase. For the interaction vertex describing superconducting pair coupling with the experimentally relevant symmetry, we find a suggestive overall agreement with the corresponding experimental curve for T_c in the cuprate $La_{2-x}Sr_xCuO_4$.
This paper is organized as follows: In the following chapter, we describe the microscopic three-band Hubbard model used for QMC simulations and weak-coupling approximations. The next two chapters present the grand canonical QMC algorithm and the fundamental fermionic sign problem. We further describe the FLEX approximation which leads to a self-consistent many-body renormalized single-particle propagator. The next chapter is dedicated to the numerical solutions of both techniques followed by results for the one-particle excitation spectrum, the magnetic susceptibility, and superconducting properties of the cuprates.

THREE-BAND HUBBARD MODEL

Our numerical investigations concerning normal and superconducting properties of the high-T_c cuprates are based on a microscopic atomic model for the two-dimensional CuO_2 planes. Ab initio band-structure calculations [14, 15], as well as spectroscopic experiments [16] clearly indicate that the physically relevant electronic states, which are characterized by low excitation energies away from the Fermi level, are constructed by hybridization of planar copper $3d_{x^2-y^2}$ and oxygen $2p_x$, $2p_y$ orbitals. The neighborhood of antiferromagnetism and superconductivity and the metal-insulator phase transition show that correlation effects play a crucial role in the high-T_c materials [17]. We model electronic correlations by a strong completely localized Coulomb repulsion between electrons which occupy the same copper site. As consequence of the Pauli exclusion principle, these electrons have opposite spin coordinates[18]. In Fig. 1, we have illustrated the topology of the CuO_2 plane including the parameters of the three-band Hubbard model [12, 13].
In second quantization, the Hamilton operator takes the following tight-binding form:

$$\begin{aligned} H &= K + V \ , \quad \text{with} \\ K &= \sum_{i,\sigma}(E_d-\mu)d^\dagger_{i\sigma}d_{i\sigma} + \sum_{j,\sigma}(E_p-\mu)p^\dagger_{j\sigma}p_{j\sigma} \\ &\quad - \sum_{<i,j>,\sigma} t_{ij}(d^\dagger_{i\sigma}p_{j\sigma} + h.c.) \ , \\ V &= U_d \sum_i d^\dagger_{i\uparrow}d_{i\uparrow}d^\dagger_{i\downarrow}d_{i\downarrow} = U_d \sum_i n_{di\uparrow}n_{di\downarrow} \ . \end{aligned} \tag{1}$$

Within the formalism of second quantization, the operators $d^\dagger_{i\sigma}$ and $p^\dagger_{j\sigma}$ create Cu $3d_{x^2-y^2}$ and O $2p_{x,y}$ electrons with spin σ on sites i respectively j, while $d_{i\sigma}$ and $p_{j\sigma}$ represent the corresponding annihilation operators. The one-electron part K is determined by the atomic energy levels at Cu sites (E_d) and at O sites (E_p), with the charge transfer energy defined by $\Delta = E_p - E_d$. The hybridization of the adjacent copper $3d_{x^2-y^2}$ and oxygen $2p_{x,y}$ orbitals is included by the hopping term $t_{ij} = \pm t$ between nearest-neighbor Cu-O

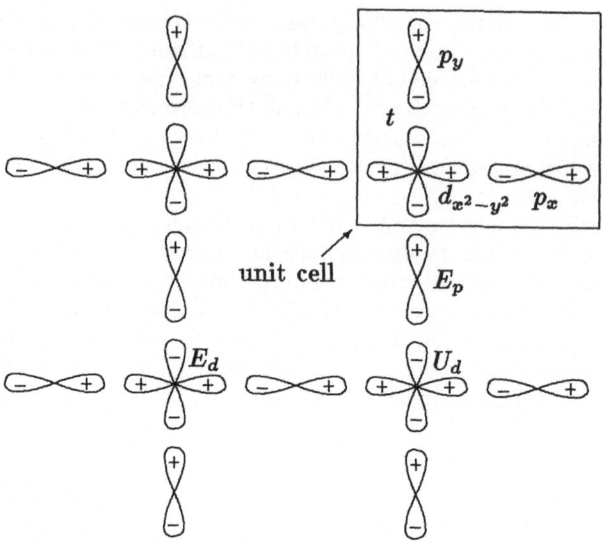

Figure 1: Relevant orbitals within the CuO_2 plane including the parameters of the three-band Hubbard model. The signs indicate the phases of the orbitals in the corresponding space segment.

sites $< i,j >$, where the sign is given by the product of the phases of the two involved orbitals as shown in Fig. 1. In the grand canonical ensemble, the chemical potential μ allows to adjust the hole doping of the system. The two-electron or interaction part V consists of the local Coulomb repulsion U_d at copper sites. Throughout this article, the hybridization parameter t, which takes the value of about 1.3 eV, is used as energy unit ($t = 1$). From recent QMC simulations [3, 4, 5], we have extracted the relevant parameter set $U_d = 6$, $\Delta = 4$ of the three-band Hubbard model, which leads to a close agreement with experimental data for a variety of normal-state and superconducting properties of the cuprate materials. In the remainder of the article this parameter set is used for QMC simulations and weak-coupling approximations.

QUANTUM MONTE CARLO SIMULATION

In this section, we review the basic notions of the Quantum Monte Carlo (QMC) algorithm used to perform finite temperature simulations of interacting electrons on a lattice, working in the grand-canonical ensemble. The application of this technique to the one-band Hubbard model was first proposed by Blankenbecler, Scalapino, and Sugar [19], and by Hirsch [20]. Our aim is to calculate the expectation value of a physical observable A, defined by:

$$< A >= \frac{Tr\left(Ae^{-\beta H}\right)}{Tr\left(e^{-\beta H}\right)} , \quad (2)$$

where $\beta = 1/T$ is the inverse temperature and the trace Tr extends over the basis of the physical (Fock) space. We now concentrate on the denominator of Eq. 2 which defines the grand partition function Z. The inverse temperature interval β is divided into L small imaginary time steps $\Delta\tau$ as $\beta = L \cdot \Delta\tau$. Applying the Trotter formula [21, 22], the grand partition function can now approximately be decomposed:

$$Z = Tr\left(e^{-L\cdot\Delta\tau H}\right) \approx Tr\left(e^{-\Delta\tau V}\, e^{-\Delta\tau K}\right)^L , \quad (3)$$

where $H = K + V$ according to Eq. 1. Since the two parts of the total Hamilton operator H do not commute, $[K, V] \neq 0$, a systematic error of order $(\Delta\tau)^2$ has been introduced here, which can be controlled at constant β by the total number of time steps L. The two-dimensional lattice model given by the above Hamilton operator H is effectively transferred to a (2+1)-dimensional discrete system including space and imaginary time coordinates.

After applying Trotter's formula to separate the one- and two-electron parts, we now bring the interaction term V into a bilinear form which couples to a bosonic auxiliary field. The so-called discrete Hubbard-Stratonovich transformation [23, 24] for a local Coulomb repulsion V is given by:

$$\exp\left(-\Delta\tau U_d n_{di\uparrow} n_{di\downarrow}\right) = \frac{1}{2} \sum_{s_{i,l}=\pm 1} \exp\left[\lambda s_{i,l}(n_{di\uparrow} - n_{di\downarrow}) - \frac{1}{2} U_d (n_{di\uparrow} + n_{di\downarrow})\right] , \quad (4)$$

which is carried out at each copper site i and for each imaginary time slice l. The constant λ is defined by the relation $\tanh^2(\lambda/2) = \tanh(\Delta\tau U_d/4)$. The Hubbard-Stratonovich transformation Eq. 4 reduces the interaction part V consisting of four fermionic creation and annihilation operators to a quadratic term of these operators coupled to a bosonic, spin-like, auxiliary field $s_{i,l} = \pm 1$. This spin-like field $s_{i,l}$ now mediates the repulsive Coulomb interaction between the electrons of the system.

By the application of the Trotter formula and the discrete Hubbard-Stratonovich transformation, we have mapped the original two-dimensional quantum system onto a (2+1)-dimensional classical (Ising) system which now can be treated with classical Monte Carlo algorithms [25]. The integration of the fermions can be performed by calculating the corresponding fermion determinants. We only state the final result for the grand partition function, while the details of the calculations are beyond the scope of this article and can be found elsewhere [7, 26]:

$$Z = \sum_{\{s_{i,l}=\pm 1\}} \left[det M^+(s) det M^-(s)\right] , \quad (5)$$

where

$$M^\pm = I + B_L^\pm \cdot B_{L-1}^\pm \cdot \ldots \cdot B_1^\pm , \quad (6)$$

and

$$B_l^\pm = e^{v^\pm(l)} \cdot e^{-\Delta\tau k} . \quad (7)$$

The unit matrix was denoted by I, and k is the matrix representation of the operator K. In the last line, we used the abbreviation:

$$\begin{aligned} v^\pm(l)_{i,i'} &= \delta_{i,i'}\left[\pm\lambda s_{i,l} + \Delta\tau\left(\mu - E_d - \frac{U_d}{2}\right)\right] , \\ v^\pm(l)_{j,j'} &= \delta_{j,j'}\left[\Delta\tau\left(\mu - E_p\right)\right] . \end{aligned} \quad (8)$$

The indices i, i' refer to copper sites, and j, j' to oxygen sites. A sequence of Hubbard-Stratonovich fields $\{s_{i,l}\}$, which are distributed in configuration space according to the probability $w = detM^+ \cdot detM^-/Z$, is generated using a standard Monte Carlo method [25].
Expectation values of physical observables A can usually be expressed in terms of Green's functions for the electrons which move within the fluctuating spin-like field by using Wick's theorem [27]. For this case, we can derive expressions similar to the above equations for the grand partition function Z. For the equal-time Green's function one achieves the following result:

$$G_{ij}^{\sigma}(\tau,\tau) = \frac{Tr\left(c_{i\sigma}(\tau)\, c_{j\sigma}^{\dagger}(\tau)\, e^{-\beta H}\right)}{Z} = \left[\left(I + B_l^{\sigma} \cdots B_1^{\sigma} B_L^{\sigma} \cdots B_{l+1}^{\sigma}\right)^{-1}\right]_{ij} . \tag{9}$$

THE SIGN PROBLEM

For fermionic systems, as the three-band Hubbard model, with repulsive Coulomb interactions, the determinants $detM^{\pm}(s)$ of a given spin configuration $\{s_{i,l}\}$ is no longer positive definite. In this case, we use the absolute value of w for definition of the statistical weight:

$$w_a[s_{i,l}] := |w[s_{i,l}]| . \tag{10}$$

By separation of the original "probability distribution" into its absolute value times its sign, $w[s_{i,l}] = sign\,(w[s_{i,l}]) \cdot w_a[s_{i,l}]$, the expectation value for a physical observable A is given by the exact identity:

$$\begin{aligned} < A > &= \frac{1}{Z} \cdot Tr\left(A\, e^{-\beta H}\right) \\ &= \frac{\sum_{\{s_{i,l}\}} A\; sign\,(w[s_{i,l}])\, w_a[s_{i,l}]}{\sum_{\{s_{i,l}\}} sign\,(w[s_{i,l}])\, w_a[s_{i,l}]} \\ &= \frac{< A \cdot sign >_{w_a}}{< sign >_{w_a}} . \end{aligned} \tag{11}$$

The symbol $< \; >_{w_a}$ denotes an average value using the absolute value of the fermionic determinants as probability distribution w_a. If the average sign given by the denominator $< sign >_{w_a}$ of Eq. 11 becomes very small, the simulation will be completely deteriorated by large fluctuations. This causes the so-called "sign problem" [7, 8] which particularly appears if $< sign >_{w_a}$ is smaller than the corresponding statistical error of the Monte Carlo simulation. The average sign obeys the following empirical form [8]:

$$< sign >_{w_a} \sim e^{-aN/T} , \tag{12}$$

where a is a constant depending on U_d and $< n_e >$, and N is the total number of lattice sites. For low temperatures and for the simulation of large systems, the average sign exponentially tends towards zero, which imposes severe constraints for Monte Carlo simulations of the Hubbard model. The appearance of negative fermionic determinants, in every case, causes a larger number of CPU-time consuming Monte Carlo measurements in order to obtain reliable numerical data. Therefore, the study of the "sign-problem" and the search for better conditioned simulation algorithms are widely discussed problems in the field of QMC simulations of interacting electrons [28].

PROPAGATOR RENORMALIZED WEAK-COUPLING APPROXIMATION

In order to obtain numerical results for larger lattices and lower system temperatures, we also applied more traditional diagrammatic perturbation expansions. Within the framework of a weak-coupling approach, the two-particle interaction term V is treated as a "small" quantity compared to the one-electron part K. For realistic model parameters which are determined by recent QMC studies [3, 4, 5] in agreement with experimental data for the high-T_c materials, however, the ratio of both parts of the Hamilton operator, Eq. 1, is a magnitude of order one. Therefore, a weak-coupling approximation of the three-band Hubbard model in the relevant parameter regime needs a careful validation by comparison with other quantitative techniques, such as the QMC simulation. For lower temperatures and electronic densities at which the Monte Carlo algorithms are plagued by the sign-problem, on the other hand, the self-consistent numerical solution of weak-coupling equations may provide further meaningful supplementations to the, in principle, exact simulations.

We apply the so-called fluctuation exchange approximation (FLEX) which was recently developed by Bickers and Scalapino [9] for the one-band Hubbard model taking into account the interaction of electrons with particle-hole fluctuations (i.e. density and magnetic correlations) and particle-particle fluctuations (i.e. superconducting pair-correlations). Within the framework of this theory, the corresponding instabilities of the normal paramagnetic phase yield estimations of transition temperatures. We now shortly sketch the ideas of the FLEX approximation applied to the three-band Hubbard model [10, 11].

To perform a diagrammatic perturbative expansion, it is useful to change the representation by a Fourier transformation of the fermionic operators:

$$\begin{aligned} a_{i\nu\sigma} &= \frac{1}{\sqrt{N}} \sum_{\vec{k}} a_{\vec{k}\nu\sigma} e^{+i\vec{k}(\vec{R}_i+\vec{\tau}_\nu)} \,, \\ a^\dagger_{i\nu\sigma} &= \frac{1}{\sqrt{N}} \sum_{\vec{k}} a^\dagger_{\vec{k}\nu\sigma} e^{-i\vec{k}(\vec{R}_i+\vec{\tau}_\nu)} \,, \end{aligned} \tag{13}$$

where index ν denotes the three orbitals ($3d_{x^2-y^2}$, $2p_{x,y}$) in the i-th unit-cell of the CuO_2 plane, and σ is the spin coordinate. The noninteracting part K of the Hamilton operator now takes the form:

$$H_0 = \sum_{\vec{k},\sigma} \sum_{\nu\nu'} [H_0(\vec{k})]_{\nu\nu'}\, a^\dagger_{\vec{k}\nu\sigma} a_{\vec{k}\nu'\sigma} \,. \tag{14}$$

Diagonalization of the matrix $[H_0(\vec{k})]_{\nu\nu'}$, which is built up by the parameters E_d, E_p, and t of the three-band Hubbard model, leads to three unperturbed energy bands $E^0_{n\vec{k}}$ labeled by the band-index n. The corresponding eigenvectors $c_{n\nu}(\vec{k})$ permit a mixed $\vec{k}$-space orbital representation of the noninteracting single-particle propagator G^0 at finite temperature T:

$$[G^0(\vec{k},\omega_m)]_{\nu\sigma,\nu'\sigma'} = \sum_n \frac{c_{n\nu}(\vec{k})c^*_{n\nu'}(\vec{k})}{i\omega_m - E^0_{n\vec{k}} + \mu} \cdot \delta_{\sigma\sigma'} \,. \tag{15}$$

Here ω_m denote fermionic Matsubara frequencies:

$$\omega_m = (2m+1)\pi T\ , \quad m = \{..., -2, -1, 0, 1, 2, ...\}\ .$$

The two-electron interaction V, which is treated as a finite-temperature many-body perturbation of this non-interacting formalism, leads to a electronic self-energy Σ given by:

$$\Sigma_\sigma^{FLEX} = \Sigma_\sigma^{HF} + \Sigma_\sigma^{(2)} + \Sigma_\sigma^{(ph)} + \Sigma_\sigma^{(pp)}, \tag{16}$$

with

$$\begin{aligned}
\Sigma_\sigma^{HF}(\vec{k},\omega_m) &= U_d\ n_{d,-\sigma},\\
\Sigma_\sigma^{(2)}(\vec{k},\omega_m) &= \frac{U_d^2}{\beta N}\sum_{\vec{q},\nu_n} G_{dd}(\vec{k}-\vec{q},\omega_m-\nu_n)\,\chi_{ph}(\vec{q},\nu_n),\\
\Sigma_\sigma^{(ph)}(\vec{k},\omega_m) &= \frac{U_d^2}{\beta N}\sum_{\vec{q},\nu_n} G_{dd}(\vec{k}-\vec{q},\omega_m-\nu_n)\,\chi_{ph}(\vec{q},\nu_n)\\
&\quad\times\left[\frac{1}{1-U_d^2\chi_{ph}^2(\vec{q},\nu_n)}-1+\frac{1}{1-U_d\chi_{ph}(\vec{q},\nu_n)}-1\right],\\
\Sigma_\sigma^{(pp)}(\vec{k},\omega_m) &= -\frac{U_d^2}{\beta N}\sum_{\vec{q},\nu_n} G_{dd}(-\vec{k}+\vec{q},-\omega_m+\nu_n)\,\chi_{pp}(\vec{q},\nu_n)\\
&\quad\times\left[\frac{1}{1+U_d\chi_{pp}(\vec{q},\nu_n)}-1\right],
\end{aligned} \tag{17}$$

where the uncorrelated particle-hole and particle-particle correlation functions are:

$$\begin{aligned}
\chi_{ph}(\vec{q},\nu_n) &= \frac{-1}{\beta N}\sum_{\vec{k},\omega_m} G_{dd}(\vec{k}+\vec{q},\omega_m+\nu_n)\,G_{dd}(\vec{k},\omega_m),\\
\chi_{pp}(\vec{q},\nu_n) &= \frac{1}{\beta N}\sum_{\vec{k},\omega_m} G_{dd}(\vec{k}+\vec{q},\omega_m+\nu_n)\,G_{dd}(-\vec{k},-\omega_m).
\end{aligned} \tag{18}$$

In Eqs. 16 - 18, we have separated different orders of the FLEX approximation. The first order, so-called Hartree-Fock term Σ^{HF} describes the interaction of electrons with the average electronic density at copper sites $n_{d,-\sigma}$ having opposite spin. In order to prevent double-counting, we have separately calculated the second order term $\Sigma^{(2)}$ and subtracted it from the particle-hole fluctuation term, $\Sigma^{(ph)}$, and particle-particle fluctuation term, $\Sigma^{(pp)}$. The full interacting Green's function G is related to the self-energy by a non-linear integral equation, the so-called Dyson's equation [27]:

$$\left[G(\vec{k},\omega_m)\right]^{-1} = \left[G^0(\vec{k},\omega_m)\right]^{-1} - \Sigma^{FLEX}(\vec{k},\omega_m)\ , \tag{19}$$

where $\left[G(\vec{k},\omega_m)\right]$ and $\left[G^0(\vec{k},\omega_m)\right]$ are matrix expressions with respect to orbital and spin indices. The self-energy Σ^{FLEX}, in return, is a functional of the full interacting Green's function G and the Coulomb repulsion U_d, as can be seen from the above Eqs. 16 - 18. All these expressions have to be solved self-consistently by adjusting the chemical potential μ to yield fixed average density of electrons per unit-cell:

$$n(T,\mu) = \frac{2T}{N}\sum_{\vec{k},\omega_m} e^{i\omega_m 0^+}\sum_{\nu=d,p_x,p_y} G_{\nu\nu}(\vec{k},\omega_m)\ . \tag{20}$$

As in the QMC simulations, the resulting full interacting Green's function G can be used to calculate several physical quantities, such as average electronic density, free-energy, or electronic excitation spectra.

DETAILS OF NUMERICAL SOLUTIONS

Quantum Monte Carlo Simulations

In order to carry out the Monte-Carlo algorithm, it is necessary to obtain the probability ratio

$$R = R_+ \cdot R_- = \frac{\omega'}{\omega} = \frac{detM^{+'} \cdot detM^{-'}}{detM^+ \cdot detM^-}, \tag{21}$$

where $M^\pm = I + B_L^\pm \cdots B_1^\pm$ according to Eq. 6. The ratio R is needed for each single spin-flip of the bosonic Ising field $\{s_{i,l}\}$, and therefore this direct calculation leads to an enormous numerical effort due to the multiplication of the large B_l^σ matrices. The following equations provide a powerful algorithm to reduce the total number of operations:

The change of the matrix B_l^σ caused by a single spin-flip $s_{i,l} \rightarrow s'_{i,l} = -s_{i,l}$ can be written as

$$B_l^\sigma \rightarrow B_l^{\sigma\prime} = (I + \Delta_i^\sigma(l))\, B_l^\sigma, \tag{22}$$

where the matrix $\Delta_i^\sigma(l)$ has only one nonzero element

$$\left(\Delta_i^\pm(l)\right)_{ii} = e^{\mp 2\lambda s_{i,l}} - 1. \tag{23}$$

With this notation the probability ratio to accept the spin-flip $s_{i,l} \rightarrow s'_{i,l}$ can be expressed by the simple equation

$$R^\sigma = \frac{detM^{\sigma\prime}}{detM^\sigma} = 1 + [1 - G_{ii}^\sigma(l,l)] \left(\Delta_i^\pm(l)\right)_{ii}. \tag{24}$$

Thus the calculation of the determinant ratio is trivial to carry out if one has already evaluated the equal-time Green's function. Using any standard algorithm, such as heat bath or Metropolis, one can decide whether a spin-flip is accepted or not. The Green's function is also needed to get the expectation value of a physical observable. If the spin $s_{i,l}$ is flipped, the new Green's function can evaluated through the relation

$$G^\sigma(l,l)' = G^\sigma(l,l) - \frac{G^\sigma(l,l)\Delta_i^\sigma(l)\, [I - G^\sigma(l,l)]}{1 + [1 - G_{ii}^\sigma(l,l)]}. \tag{25}$$

Since $\Delta_i^\sigma(l)$ has only one nonzero element, it takes N^2 operations to update the equal-time Green's function. Contrary to that the direct calculation according to Eq. 9 would need an effort of the order of $\mathcal{O}(L \cdot N^3)$ for each spin-flip. After updating all Ising spins of a single time-slice, the following equation allows to "wrap" the Green's function to the next time-slice:

$$G^\sigma(l+1, l+1) = B_{l+1}^\sigma G^\sigma(l,l) \left(B_{l+1}^\sigma\right)^{-1}. \tag{26}$$

Unfortunately the updating described by Eq. 25 is numerical unstable. Thus it is necessary to evaluate the Green's function exactly from Eq. 9 after several time steps. A clever storage of the intermediate products of the B_l^σ matrices during the calculation of $G^\sigma(l,l)$ allows to reduce this effort to a minimum level. Altogether the CPU-time for one Monte-Carlo sweep over all lattice sites and time-slices is of the order $\mathcal{O}(L \cdot N^3)$. A typical Monte-Carlo run consists of at least 1000 measurements after several warm up sweeps. In order to assure statistical independence it is necessary to run a few sweeps between each measurement.

Fluctuation Exchange Approximation

The self-consistent solution of the nonlinear integral equation (Eq. 19) is performed on a discrete mesh in lattice momentum and frequency space. At finite temperature T, the Matsubara frequencies ω_m provide a natural discretization of frequency sums which are limited by a high-energy cutoff ω_c. For numerical frequency summations, which appear in the particle-hole and particle-particle correlation functions (Eqs. 18), and self-energy (Eqs. 17), a fixed sharp frequency cutoff ($\omega_c = 50t$) has been chosen in order to prevent artificial temperature dependencies [29]. This procedure demands a doubling of the number of Matsubara frequencies and setting the Green's function to zero between the cutoff and twice the cutoff. Furthermore, the total number of Matsubara frequencies linearly increases with the inverse temperature β (160 frequencies at $T = 0.1t$). For integrations in $\vec{k}$-space, we have divided the Brillouin zone (i.e. the unit cell in $\vec{k}$-space) into a discrete mesh of equidistant points, which is formally equivalent to a finite real-space lattice with periodic boundary conditions.
A considerable reduction of storage size and computing effort can be achieved by observing the lattice symmetries: The self-energy is calculated only inside the irreducible part, which extends over one-eighth of the Brillouin zone, and subsequently obtained in the full zone by symmetry operations. Nevertheless, at each stage of the iterative process, three summations over $\vec{k}$-vectors and Matsubara frequencies have to be performed in order to calculate the uncorrelated particle-hole and particle-particle susceptibilities (Eqs. 18) and the self-energy (Eqs. 17), which causes most of the computing effort of the self-consistent algorithm.
We start the iteration with an initial approximation for the Green's function $G = G^0(\mu_0)$, where the chemical potential μ_0 is chosen to give the required electronic density in the absence of the two-particle interaction U_d. The density $n(T,\mu)$ is computed after each iteration step, and μ is adjusted using Newton's method to reduce the difference between the computed density and the selected density. During the self-consistent process, we have to insure that the denominators in Eqs. 17 do not vanish or become negative. In order to achieve a stable convergence, the Coulomb repulsion parameter U_d is increased slowly ("adiabatically") to its full magnitude starting from small value. We employ, as commonly used within iterative calculation, a feedback after each iteration step, i.e. we retain a weighted average of the self-energy from two successive iterations, to avoid convergence oscillations. The calculation is stopped, if the following convergence criteria are fulfilled: (1) the calculated density differs from the selected density by less than 0.001. (2) The relative magnitude of the one-electron propagator (the full Green's function G) computed on two successive iterations differs less than 0.1 percent for all $(\vec{k}, \omega_m)$ points.

We now compare numerical results for the one-electron low-energy excitation spectrum, which have been derived from QMC simulations and self-consistent solutions of the FLEX equations, with experimental data. Angle-resolved photoemission (ARPES) and inverse photoemission (ARIPES) measurements [30] clearly show the existence of a dispersive energy band $E(\vec{k})$ crossing the Fermi-level E_F. In the grand canonical QMC technique, as well as in the weak-coupling FLEX approximation, we calculate the thermodynamic Green's function G at imaginary time or frequency coordinates. In order to evaluate the spectral density as function of real frequency ω we have to perform an analytic continuation. For the "noisy" QMC data, we used a modified least-square fit [31] to obtain the spectral density $A(\vec{k},\omega)$:

$$A(\vec{k},\omega) = -\frac{1}{\pi} \cdot \mathrm{Im}G(\vec{k},\omega) \; , \tag{27}$$

while, for the FLEX approximation, we applied a standard Padé algorithm [32] for analytic continuation of G. For each $\vec{k}$-vector, we have determined the maximum of $A(\vec{k},\omega)$ as function of the real frequency ω closest to the Fermi-level E_F. In Fig. 2, we show the

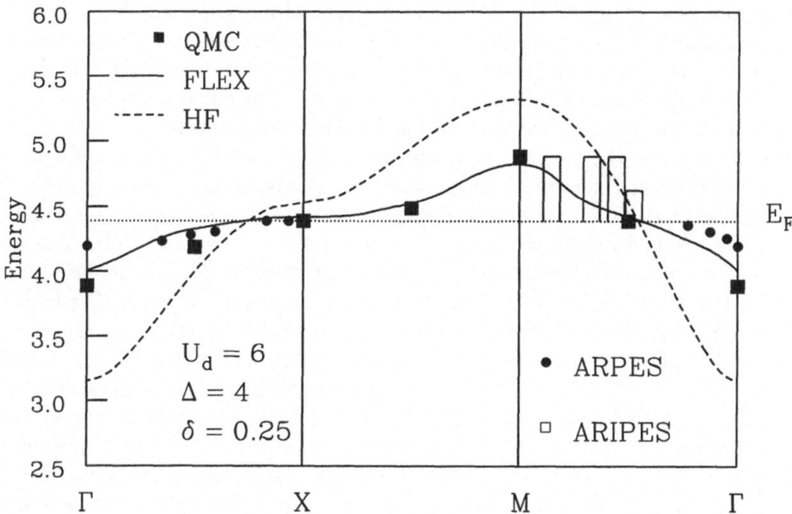

Figure 2: Dispersion $E(\vec{k})$ of the correlated electronic states near the Fermi-level E_F derived from QMC simulations, fluctuation exchange approximation (FLEX), and mean-field Hartree-Fock (HF) calculation. Experimental data from spectroscopic measurements [30] (ARPES: angle resolved photoemission, ARIPES: angle resolved inverse photoemission) are provided additionally.

corresponding energy $E(\vec{k})$ for a hole doping $\delta = 0.25$ of the CuO_2 planes at the inverse temperature $\beta = 10$. The dispersion of the correlated electronic states near the Fermi level calculated from QMC and FLEX is in good agreement with ARPES and ARIPES experimental data [30] which were measured in the cuprate $Bi_2Sr_2CaCu_2O_8$ at a doping concentration of about 0.20. This result shows that the three-band Hubbard model accurately describes the low-energy single-particle excitations of the superconducting oxides.

The simpler weak-coupling Hartree-Fock (HF) approximation, which completely neglects two-particle fluctuation effects, leads to a bandwidth and Fermi velocity (i.e. the slope of $E(\vec{k})$ at E_F) quite different from the experimental values. This result indicates that electronic correlations, which are included in both the QMC and FLEX calculations, renormalize considerably the single-particle excitation spectrum compared to a purely one-particle treatment (HF).

MAGNETIC SUSCEPTIBILITY

The static ($\omega = 0$) magnetic susceptibility is defined by:

$$\chi(\vec{q}, \omega = 0) = \frac{1}{N} \sum_{i,i'} e^{i\vec{q}\cdot(\vec{R}_i - \vec{R}_{i'})} \int_0^\beta d\tau < (n_{i\uparrow}(\tau) - n_{i\downarrow}(\tau)) (n_{i\uparrow}(0) - n_{i\downarrow}(0)) > , \qquad (28)$$

where the indices i and i' only denote Cu sites. We calculate this quantity, which provides insights to the magnetic ordering of the electronic system, using the QMC technique and weak-coupling approximations. Within the framework of QMC simulations the two-particle expectation value in Eq. 28 is directly measured, while, for the weak-coupling analysis, we apply the so-called "parquet" approximation [33, 34]. This perturbative diagrammatic approach treats both the one-particle self-energy and the two-particle vertex functions, which are necessary to compute susceptibilities like the above defined $\chi(\vec{q}, \omega = 0)$, in a symmetric self-consistent formalism. The corresponding analytic parquet expressions, however, are beyond the scope of this review article.

In Fig. 3 we show the magnetic susceptibility $\chi(\vec{q}, \omega = 0)$ for the commensurate wave vector $\vec{q} = (\pi, \pi)$ as function of temperature T. A divergence of this quantity indicates long-range antiferromagnetic spin ordering of the electronic system. Calculations were performed for hole doping $\delta = 0.25$ on a lattice of 4×4 CuO_2 unit cells. The results of a simple mean-field Hartree-Fock (HF) analysis suggest an antiferromagnetic instability at very high temperatures ($T_{AF} \approx t$) far away from agreement with QMC simulations and experimental data. Neutron scattering measurements [35, 36, 37] show the existence of a non-ordered paramagnetic phase for the cuprate materials at high hole concentrations with remaining short-range incommensurate spin fluctuations near the antiferromagnetic wave vector. In our QMC and parquet calculations these short-range magnetic fluctuations lead to a enhanced susceptibility at low temperatures. The electronic system, however, does not undergo an antiferromagnetic phase transition down to very low temperatures ($T = 0.02t$) as can be seen from weak-coupling parquet calculations. These temperatures, in contrast, cannot be reached by QMC simulations due to sign problems.

SUPERCONDUCTING PROPERTIES

Finally, we discuss the problem of a attractive pairing interaction, which may lead to superconductivity, within the framework of a microscopic repulsive Hubbard model. In order to show an attractive coupling for the three-band Hubbard model, we consider the equal time pairing correlation function for the extended s-wave symmetry of the order parameter:

$$P_{s^*} = < \Delta_{s^*} \Delta_{s^*}^\dagger > , \qquad (29)$$

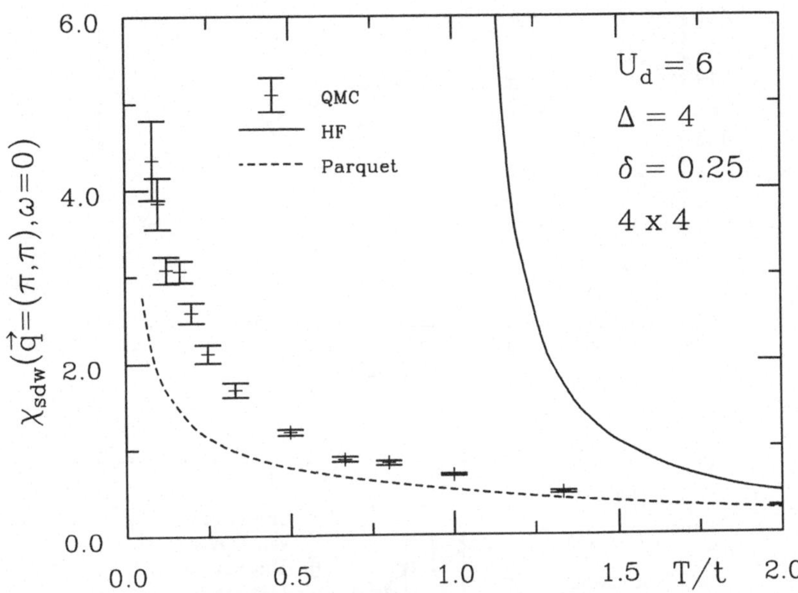

Figure 3: Static ($\omega = 0$) magnetic susceptibility as function of temperature at the commensurate wave vector $\vec{q} = (\pi, \pi)$ describing antiferromagnetic spin ordering of the electronic system. Results of QMC simulations are compared with weak-coupling data employing the self-consistent parquet and Hartree-Fock approximation

where $\Delta^{\dagger}_{s^*}$ denotes a pair-field operator creating pairs of particles only at neighboring lattice sites as consequence of the short-range Coulomb repulsion. The operator Δ_{s^*} is the corresponding annihilation operator. For an explicit definition of the underlaying operators see Ref. [38]. We now define the interaction vertex Γ_s as the difference between the correlation function P_{s^*} and the corresponding quantity for two one-particle excitations that propagate without interaction. A positive vertex means that the interaction enhances pairing. In Fig. 4(a), we have displayed QMC data for Γ_s as function of hole doping δ of the CuO_2 planes. The simulations have been performed on a 4×4 lattice at the inverse temperature $\beta = 10$. It can be seen as a remarkable result that an attractive pairing interaction induced by electronic correlation arises from a purely repulsive microscopic model, i.e. the three-band Hubbard model. The maximum appears for a doping of $\delta \approx 0.20$, and the overall shape of the curve is very reminiscent of the corresponding experimental curve [39] for superconducting transition temperature T_c in the cuprate $La_{2-x}Sr_xCuO_4$ shown in Fig. 4(b). In order to proof long-range order of the electronic system, the interaction vertex Γ_s has been studied as function of the lattice size. The QMC simulations have not (yet) shown the existence of a macroscopic superconducting phase. This may be due to the fact that the relevant temperature scale for superconductivity was still not reached for the extended s-wave symmetry with the parameter set used in the present model. Again, sign problems are limiting the possibility to obtain QMC data for lower temperature. Nevertheless, Fig. 4 implies that the three-band model remains a serious candidate for superconductivity induced by electronic correlation in the experimentally relevant symmetry channel of the order parameter.

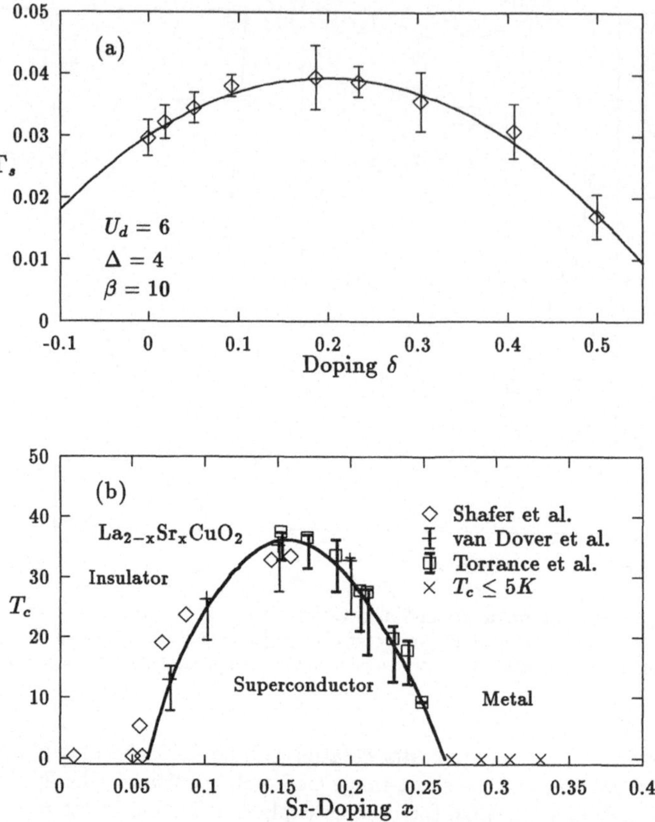

Figure 4: (a) Coupling strength of the attractive pairing interaction (i.e. the interaction vertex Γ_s for extended s-wave symmetry) for the three-band Hubbard model as function of hole doping. The QMC-simulation was performed for 4×4 CuO_2 unit cells. (b) The superconducting transition temperature T_c of the cuprate $La_{2-x}Sr_xCuO_4$ measured experimentally as function of the Sr content x which equals the hole doping δ of the CuO_2 planes. Experimental results are taken from Ref. [39].

CONCLUSIONS

In conclusion, we have presented two different theoretical approaches, i.e. Quantum Monte Carlo simulations and the perturbative fluctuation exchange approximation, to the three-band Hubbard model for the high temperature superconductors leading to very accurate numerical results for the low-energy single-particle excitations. The experimentally observed Fermi surface and Fermi velocity is reproduced quantitatively, in contrast to mean-field based Hartree-Fock and local density approximations. Calculations of the magnetic susceptibility for finite hole doping show the existence of short-range incommensurate fluctuations of the electronic systems at low temperatures. In contrast to simpler mean-field (Hartree-Fock) estimations, which lead to an unphysical instability of the normal state at a very high temperature, the inclusion of fluctuations to the weak-coupling approach suppresses the antiferromagnetic phase transition down to very low temperatures.

Finally, the same parameter set, as has been used for the normal-state properties, yields an attractive interaction vertex in the extended s-wave channel, where the doping dependence is analogous to the one of T_c as measured experimentally.

ACKNOWLEDGEMENTS

We are grateful to the HLRZ-Jülich for providing us computing time at the Cray Y-MP/832 as part of the project "High Temperature Superconductivity". The weak-coupling calculations were performed on the Cray Y-MP EL of the Rechenzentrum der Universität Würzburg and on the Cray Y-MP/864 of the Leibniz-Rechenzentrum München under a cooperation program with Cray Research Inc. We thank these institutes for their support. Furthermore, we gratefully acknowledge support by the DFG project No. Ha 1537/5-2 and by the Bavarian High-T_c program "FORSUPRA".

REFERENCES

[1] For review articles on theoretical and experimental approaches to the high temperature superconductors see, for example: Proceedings of the International Conference, Kanazawa, Physica C **185-189** (1991).

[2] For a review article concerning QMC techniques applied to Hubbard model see: W. von der Linden, Physics Reports, **220**, 53 (1992), and references therein.

[3] G. Dopf, A. Muramatsu, W. Hanke, Phys. Rev. Lett. **68**, 353 (1992).

[4] G. Dopf, J. Wagner, P. Dieterich, A. Muramatsu, and W. Hanke, Phys. Rev. Lett. **68**, 2082 (1992).

[5] G. Dopf, A. Muramatsu, W. Hanke, Europhys. Lett. **17**, 559 (1992).

[6] R. T. Scalettar et al. Phys. Rev. B **44**, 770 (1991).

[7] S. R. White et al., Phys. Rev. B **40**, 506 (1989).

[8] E. Y. Loh et al., Phys. Rev. B **41**, 9301 (1990).

[9] N. E. Bickers and D. J. Scalapino, Annals of Physics **193**, 207 (1989).

[10] R. Putz, P. Dieterich, and W. Hanke, Helv. Phys. Acta **65**, 433 (1992).

[11] J. Luo and N. E. Bickers, Phys. Rev. B **47**, 12153 (1993).

[12] C. M. Varma, et al., Solid State Comm. **62**, 681 (1987).

[13] V. J. Emery, Phys. Rev. Lett. **58**, 2794 (1987).

[14] A. K. McMahan, R. M. Martin, and S. Satpathy, Phys. Rev. B **38**, 6650 (1988).

[15] M. S. Hybertsen, M. Schlüter, and N. E. Christensen, Phys. Rev. B **39**, 9028 (1989).

[16] N. Nücker et al., Phys. Rev. B **37**, 5158 (1988).

[17] P. W. Anderson, Science **235**, 1196 (1987).

[18] J. Hubbard, Proc. Roy. Soc. **A276**, 238 (1963).

[19] R. Blankenbecler, D. J. Scalapino, and R. L. Sugar, Phys. Rev. D **24**, 2278 (1981).

[20] J. E. Hirsch, Phys. Rev. B **31**, 4403 (1985).

[21] H. F. Trotter, Prog. Am. Math. Soc. **10**, 545 (1959).

[22] M. Suzuki, Prog. Theo. Phys. **56**, 1454 (1976).

[23] J. Hubbard, Phys. Rev. Lett. **3**, 77 (1959).

[24] J. E. Hirsch, Phys. Rev. B **28**, 4059 (1983).

[25] For a comprehensive review see e.g.: Dieter W. Heermann, *Computer Simulation Methods in Theoretical Physics*, Springer, Berlin-Heidelberg, (1986).

[26] E. Y. Loh and J. E. Gubernatis, in: *Electronic Phase Transistions*, ed. by W. Hanke and Y. V. Kopaev, North-Holland (1992).

[27] A. L. Fetter and J. D. Walecka, *Quantum-Theory of Many Particle Systems*, New York (1971).

[28] A. Muramatsu, G. Zumbach, X. Zotos, Int. J. Mod. Phys. C **3**, 185 (1992).

[29] J. W. Serene and D. W. Hess, Phys. Rev. B **44**, 3391 (1991).

[30] G. Mante et al. Z. Phys. B **80**, 181 (1990).

[31] S. R. White et al. Phys. Rev. Lett. **63**, 1523 (1989).

[32] H. J. Vidberg and J. W. Serene, J. Low Temp. Phys. **29**, 179 (1977).

[33] C. de Dominicis and P. C. Martin, J. Math. Phys. **5**, 14 (1964); *ibid.*, 31 (1964).

[34] N. E. Bickers and S. R. White, Phys. Rev. B **43**, 8044 (1991).

[35] J. Rossat-Mignod et al., Physica B **169**, 58 (1991); Physica C **185-189**, 86 (1991).

[36] J. M. Tranquada et al., Phys. Rev. B **46**, 5561 (1992).

[37] T. E. Mason, G. Aeppli, and H. A. Mook, Phys. Rev. Lett. **68**, 1414 (1992).

[38] G. Dopf, A. Muramatsu, and W. Hanke, Phys. Rev. B **41**, 9264 (1990).

[39] J. B. Torrance et al., Phys. Rev. Lett. **61**, 1127 (1988), and references therein.

DYNAMIC SYSTEMS VISUALIZATION APPLIED TO FLIGHT MECHANICS PROBLEMS

G. Sachs
Institute of Flight Mechanics and Flight Control
Technische Universität München
Arcisstraße 21
Munich, Germany

SUMMARY

Visualization is considered as a supporting tool for investigations on dynamic systems. A technique is described for generating images of moving three-dimensional objects including possible deformations. Flight mechanics problems are used as examples of which one concerns the separation of a two-stage flight system equipped with air-breathing engines at hypersonic speed. The other example deals with dynamic soaring which is a complex flight technique for extracting energy from horizontal wind.

INTRODUCTION

The high performance of computers and graphics software now available makes it possible to visualize complex three-dimensional objects and their motions. Due to these computational capabilities, visualization may be used as a means for supporting investigations on the dynamics of systems and related control and optimisation problems. The expected increase in computer and software performance may further expand the application of visualization.

Based on experience gained so far (Refs. [1-3]), it is the purpose of this paper to describe possibilities of visualization techniques for application in system dynamics. Problems of flight mechanics are used as examples. The dynamics examples considered represent rather complex optimal control problems for which solutions are to be found such that a certain performance criterion is met. The results achieved require a significant computational effort. They are based on sophisticated mathematical methods and efficient computer programs. The mathematical methods and computer programs used (Refs. [4-6]) have shown their capability and efficiency for solving complex problems in many scientific and technical areas. They provide results with high accuracy.

VISUALIZATION GOALS

Complex three-dimensional objects, their deformation and their motion in translational and rotational degrees of freedom can be graphically generated and displayed with the use of computers and, thus, complex motions can be visualized. This possibility may be utilised as a means for supporting investigations on control problems of dynamic systems. Visualization may contribute in various ways:

- Better understanding of dynamics problems when investigating complex motions
- Contributing to finding a solution
- Better communication with cooperating scientists of other disciplines
- Contributing to improving didactics in lectures
- Better presentations of results to an audience

Each of the points will be addressed in this FORTWIHR research project which is concerned with system dynamics problems. The third point may be of particular interest for the cooperation in FORTWIHR. This is because of the interdisciplinary character of the research projects of FORTWIHR where various disciplines (mathematics and computer and engineering sciences) are working together. Visualization is aimed at a better communication and cooperation of the scientists involved the research projects. Another issue of planned visualization activities concerns real-time applications. This may be of interest in regard to man-machine interface problems as it is the case in the cockpit of an aircraft. Visualization techniques may be used to improve the information which is displayed to the pilot. Information of flight maneuvers, aircraft state etc. can be presented in a new manner so that more information may be made available and presented in a way better suited for acceptance by the pilot.

IMAGE GENERATION OF THREE-DIMENSIONAL MOVING AND DEFORMABLE OBJECTS

A visualization tool has been developed to generate images of three-dimensional objects which are moving translationally and rotationally (six degrees of freedom) and which may show deformations. The tool makes use of graphics features of the applied Silicon Graphics computers (Fig. 1). Data for describing the object, its motion and deformation are provided by the application and the data aquisition module and forwarded to the system. The geometry subsystem module is concerned with transformation matrix operations. The scan conversion subsystem performs calculations on individual pixel basis. For example, this includes calculation of screen positions and parameter values of colour and texture features. Data of the scan conversion subsystem are passed to the raster subsystem, where each pixel's bits are organized as image bit-planes, depth, stencil, texture overlay/underlay, window and frame buffer feature extension planes. Data are then passed to the display subsystem which finally sends it to digital-to-analog converters for display.

Steps in the visualization process may also include definition of surface features like diffuse and specular reflection characteristics, shininess, transparency and emission properties. Lighting features can be applied as ambient light or spot light.

Two examples are considered in this paper. They concern the motion of a hypersonic flight system which is a very fast flying vehicle and a flight technique of a more slowly

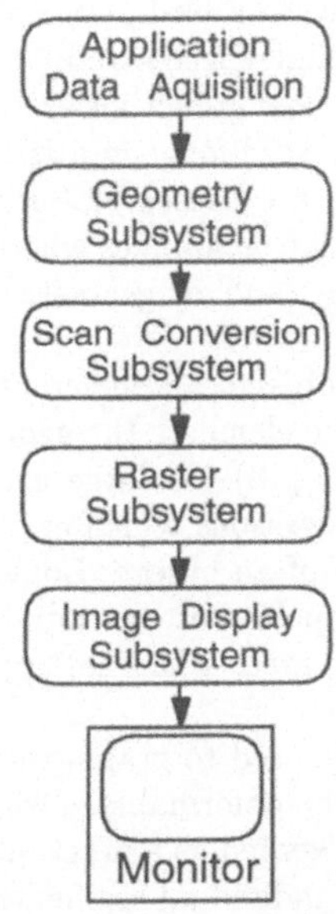

Fig. 1 Modules of graphic systems for image generation

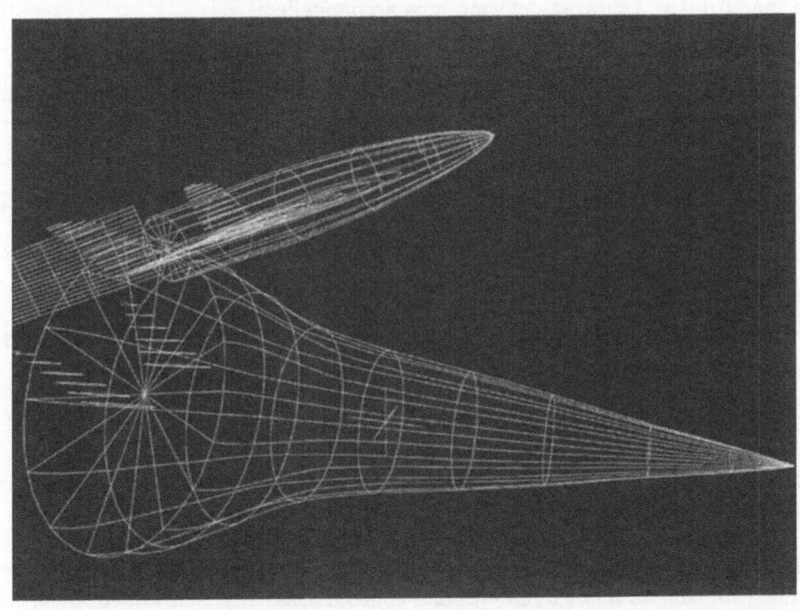

Fig. 2 Basic elements used for visualization of three–dimensional objects

flying object which may be a sailplane or a bird performing a maneuver called dynamic soaring. The flight maneuver of the high speed example is the separation of a two-stage system for releasing the upper stage to conduct an ascent to an orbit.

The described visualization tool is used for generating images of objects like airplanes and birds for which some examples are presented. Elements of these objects like a fuselage can be generated by interactive techniques where an easy handling is supported by making use of symmetry features and of geometric characteristics such as conic sections for contructing the form.

The forms of three-dimensional objects are generated with a grid as illustrated in Fig. 2. Simple forms are used as basic elements the geometry of which is changed until the final object form is achieved (Fig. 3). A large grid density is applied when the object form is highly curved. Conic sections which can be defined by three points are used for describing the cross-section of an object. Longitudinal sections are described with the use of splines. Each mesh point can be individually altered with the mouse in order to model details in a certain area. The final appearance of the objects used as examples is illustrated in Fig. 4.

The visualization tool can also be used to graphically present changes in the shape of an object when it is supposed to be deformable. Deformations are determined with computations and are graphically presented as a function of time or of another variable. The deformable parts are defined as individual segments of an object.

As an example, Fig. 5 shows flapping of the wing of a bird during flight. Another example is presented in Fig. 6 which shows the motion of a bird during take-off from the water. In addition to wing flapping and feet motion, waves due to feet dipping into the water are visualized.

For the bird, the following deformations are modelled:

- Wing
  Partitioning of wing in segments each of which can conduct a flapping, torsional and bending motion.

- Head/body
  Motion of head relative to body including opening of the bill, torsion of tail and motion of feet for running.

SEPARATION MANEUVER OF LIFTING VEHICLES AT HYPERSONIC SPEED

Currently, new systems for space transportation are considered in order to enhance performance. A promising proposal is a two-stage lifting system which is capable of an airborne flight at hypersonic speed. The first stage is equipped with an airbreathing propulsion system while the second stage is propelled by rockets. The vehicles separate at hypersonic speed and then the second stage conducts an ascent to an orbit.

The separation of two-stage lifting vehicles at hypersonic speed is a maneuver which poses new problems in flight dynamics and control. Basically, it is aimed for providing best starting conditions for the second stage for achieving an optimal ascent to an orbit. This is a problem aera related to flight performance of the vehicles. Another problem aera concerns the stability and control of the vehicles. Here, an issue of primary

Fig. 3 Grid models of three–dimensional objects

Fig. 4 Final modelling of three–dimensional objects

Fig. 5 Visualization of wing flapping

Fig. 6 Graphical simulation of waves due to dipping of feet into water

concern is the safety of the separation maneuver. An adequate control of the maneuver is required for achieving both the safety and the performance goals.

Fig. 7 shows basic elements of the separation maneuver and addresses points of primary importance. As may be seen, the overall separation maneuver may be decomposed into three phases:

- Pull-up (combined system)
- Positioning of second stage
- Separate motion of both stages

Each phase poses specific problems also indicated in Fig. 7. Control techniques capable of coping with this type of problems are required for developing a solution which is satisfactory from both a safety and performance standpoint.

With the use of visualization, the effects of different control techniques on the separation maneuver can be shown. Emphasis is put on the relative motion between the rear of the second stage and the adjacent surface of the first stage during the initial phase of separation. This is illustrated in Fig. 8. Improper control may not be able to avoid a collision. This particulary concerns the fact that a positive lift force is necessary for the second stage to move away. An increase in angle of attack for producing a positive lift force requires a rotation of the vehicle such that the rear moves downward. A positive lift change yields a correct movement of the center of gravity. However, the rear of the vehicles may initially move in the opposite direction because of the commanded rotation. Thus, a collision with the adjacent part of the first stage may be possible.

A proper control technique avoids this dangerous situation by appropriately limiting the rotation rate until a safe distance is reached. This is illustrated in Fig. 9 which shows the relative motion between the rear of the second stage and the adjacent part of the first stage and between the centers of gravity of both stages. As may be seen, it is possible to achieve a motion of the rear of the second stage such that an approaching of the first stage does not exist at any instant.

It may be of interest to note that results as shown are based on a complex mathematical description providing a realistic model and on the application of optimization techniques (Refs. [7, 8]). Thus, it is possible to develop a solution which represents an optimal separation maneuver of the vehicles, i.e., the translational displacements of the centers of gravity versus time is maximized while an unfavorable motion of the rear of the upper stage is avoided. The solution achieved shows that there is practically no performance degradation concerning the separation of both stages when compared to an unconstrained maneuver.

DYNAMIC SOARING

Dynamic soaring is a flight technique with the use of which energy can be extracted from a horizontal wind changing with altitude. This flight technique has early aquainted interest (Refs. [9-11]) and is still subject of research (Refs. [12-14]). Dynamic soaring is performed by large sea birds and is of interest also for manned flight.

The flight maneuver of dynamic soaring is rather complex when compared to flight techniques applied for extracting energy from an up-wind. In an up-wind, a bird or a sailplane is flying in a current which is moving upward. The energy transfer mechanism can be directly understood as the air lifting up the flying object. A flight maneuver for

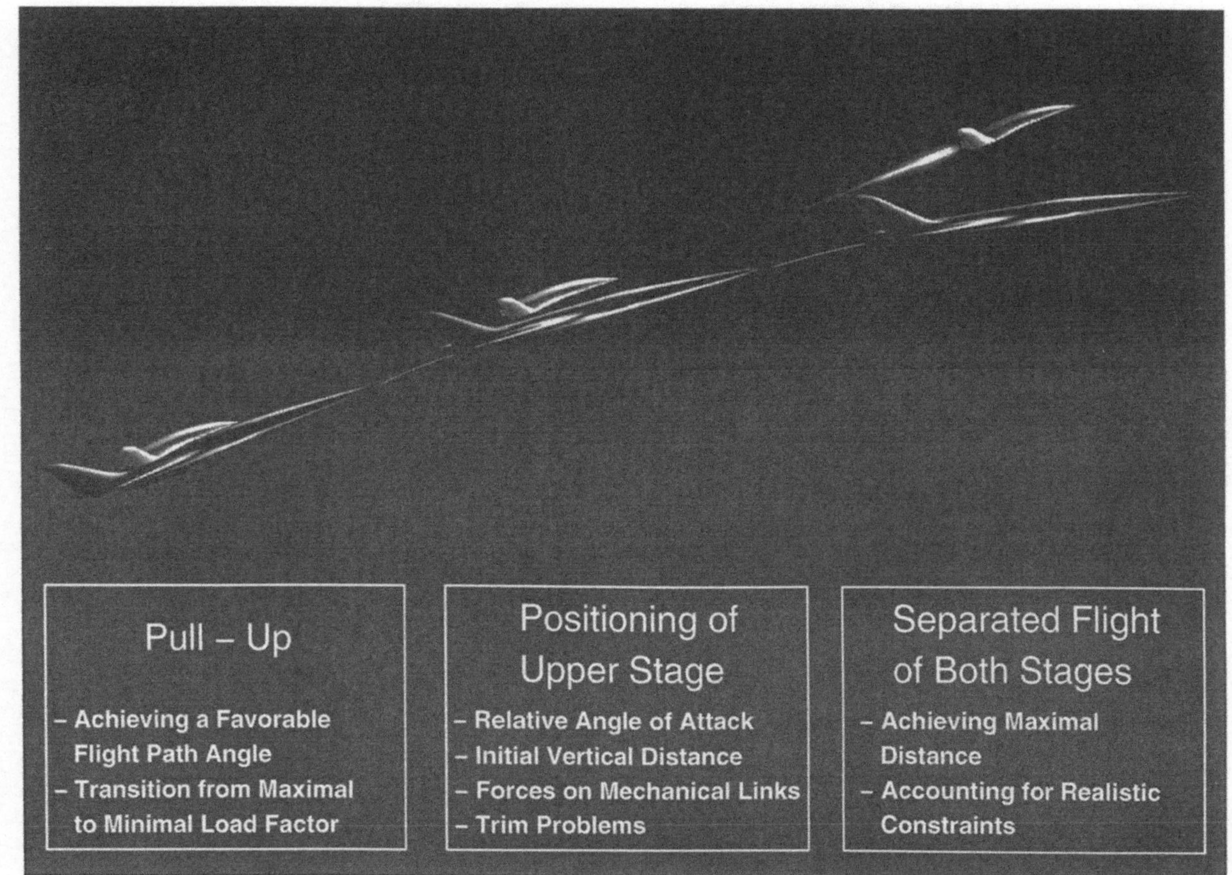

Fig. 7 Phases of the separation maneuver of hypersonic vehicles

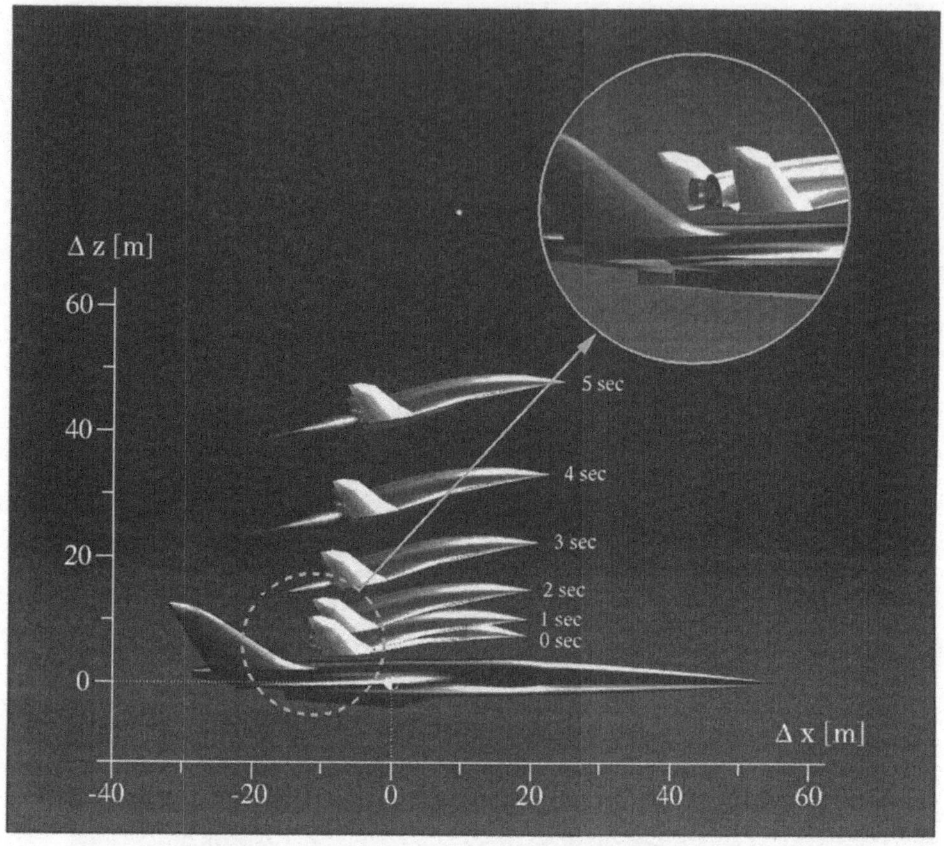

Fig. 8 Relative motion of the two stages after release of second stage

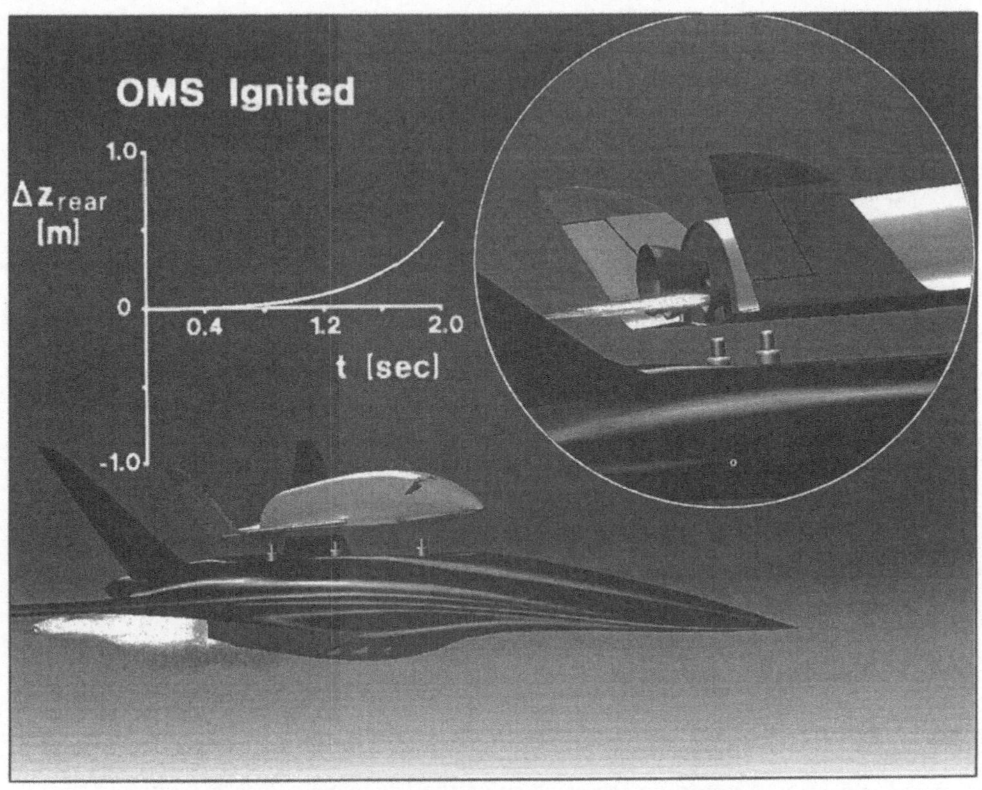

Fig. 9 Optimized maneuver after release of second stage (OMS: Orbital Maneuvering System)

utilizing this energy source can be rather straight forward, i.e., it may be a rectilinear or a circling flight. By contrast, a flight maneuver for extracting energy from a horizontal wind (shear wind) is made up of a complex combination of climbing, descending and turning flight phases. Due to its complexity, dynamic soaring may pose more difficulties in understanding the underlying physical mechanism and its unique flight profile. Visualization is considered as a means for contributing to such an understanding.

For computing dynamic soaring trajectories, sophisticated and efficient numerical simulation and optimization procedures are necessary. This particularly holds when trajectories are considered which require minimum energy extraction from the moving air. Problems of this type can be solved by applying optimal control theory and appropriate numerical procedures. Results presented in this paper make use of these techniques (Refs. [4-6]).

If may be of interest to note that the aforementioned considerations and methods contribute to the solution of an ornithology problem which concerns dynamic soaring flight of large sea birds. It is an example of interdisciplinary cooperation of mathematics, computer and engineering sciences and ornothology.

Basic characteristics of a dynamic soaring trajectory are illustrated in Fig. 10. As may be seen, the trajectory is made up of periodically repeated cycles each of which shows four phases:

- Turn close to the water surface
- Climb with head wind
- Turn at top of the trajectory
- Descent before the wind

The wind profile also illustrated in Fig. 10 shows a change with altitude. Most of this change occurs in the boundary layer which is a region close to the water surface .

The periodic trajectory depicted in Fig. 10 implies that the energy state of the bird at the end of a trajectory is the same as at the beginning. Such a trajectory may be termed energy-neutral. Among all possible energy-neutral trajectories, one exists which requires minimum wind strength for extracting energy from the moving air. An example is presented in Fig. 11 which shows a minimum wind strength trajectory of an albatros.

A physical insight into the energy transfer mechanism is provided by Fig. 12 which shows speed vectors at two points of a trajectory. The altitude values of both points agree so that they represent conditions of the same potential energy. However, kinetic energy has changed. This is due to the change in groundspeed which is an indication of kinetic energy. Groundspeed which is determined by the vector sum of the speed relative to the air and windspeed is increased in the upper part of the trajectory where a turn from head wind motion towards a motion before the wind is performed.

Shear wind conditions which also exist at higher altitude may be used by sailplanes as an energy source. This is illustrated in Fig. 13 which shows basic characteristics of a dynamic soaring trajectory of sailplanes.

CONCLUSIONS

Visualization is considered as a means for supporting investigations on dynamic systems. One problem described refers to the separation maneuver of lifting two-stage

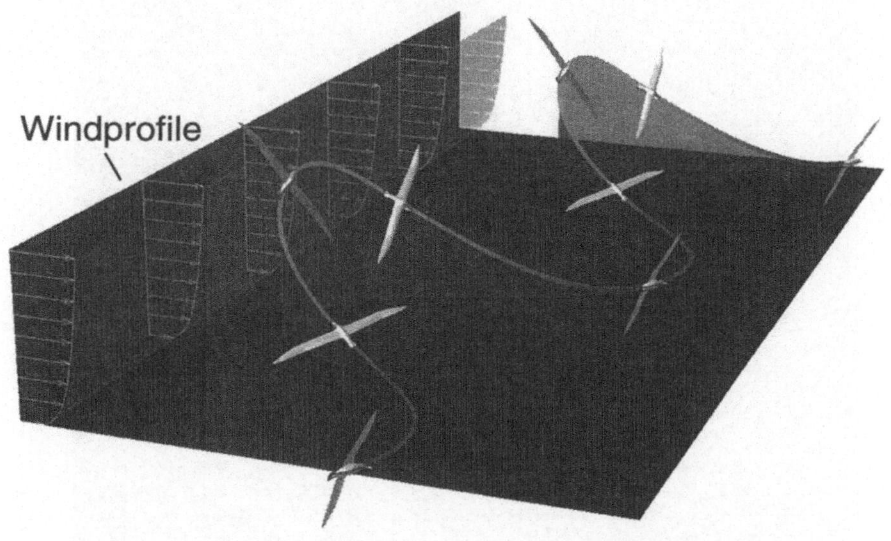

Fig. 10 Basic characteristics of dynamic soaring (different scales for trajectory and bird)

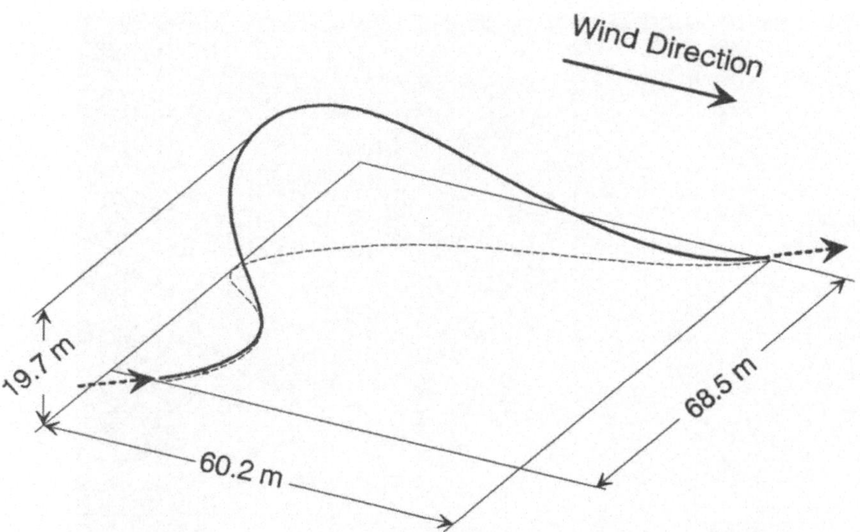

Fig. 11 Optimal trajectory requiring minimum wind strength for dynamic soaring of albatrosses

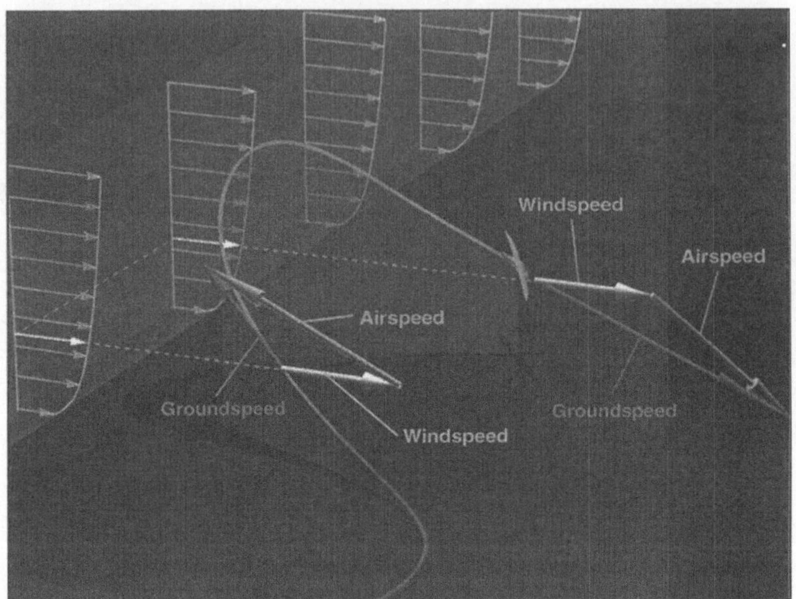

Fig. 12 Speed vectors at two trajectory points of equal altitude

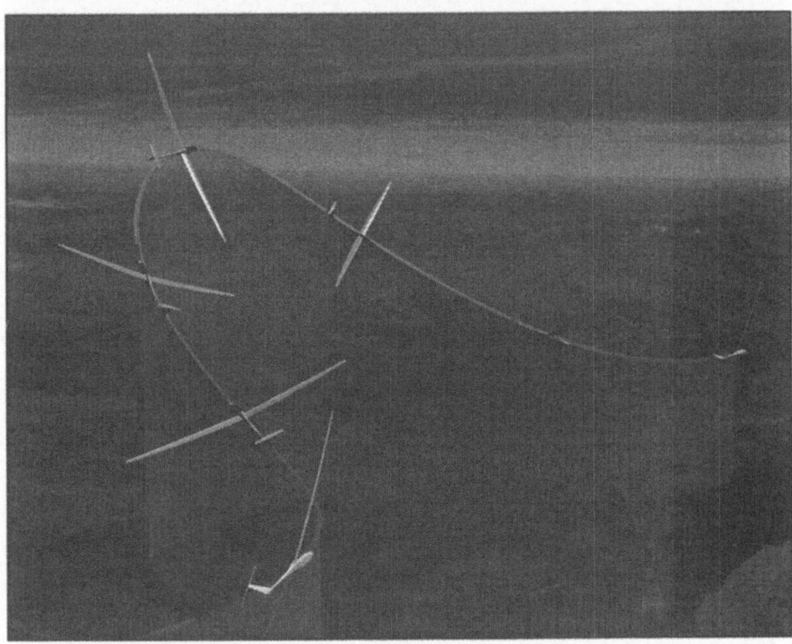

Fig. 13 Illustration of dynamic soaring of sailplanes (different scales for trajectory and sailplane)

vehicles at hypersonic speed. The other problem concerns dynamic soaring which is a flight technique performed by large sea birds and which may be also of interest for sailplanes.

A technique for generating images of three-dimensional objects is described. This includes their motion in translational and rotational degrees of freedom as well as deformations of the objects.

Details of the hypersonic flight example concern different phases of the separation maneuver. Control of the vehicles following release of the second stage is addressed, with safety considerations taken into account. Visualization is applied in order to show an optimal separation maneuver which avoids unsafe effects and yet has practically no performance degradation.

A visualization of basic characteristics of dynamic soaring is presented which is a rather complex flight maneuver for extracting energy from horizontal wind. This includes a physical insight into the mechanism how this energy transfer is accomplished by an appropriate trajectory. The visualized dynamic soaring trajectories concern the flight of large sea birds and of sailplanes.

REFERENCES

[1] Sachs, G., Möller, H. Knoll, A., Mehlhorn, R., Deicke, T.: Der Albatros als Fluglehrer. Lehrstuhl für Flugmechanik und Flugregelung der Technischen Universität München sowie Bayerischer Rundfunk/Fernsehen, München, Videofilm, Betacam SP, 1992.

[2] Sachs, G., Mehlhorn, R., Möller, H.: Visualisierung von Optimalsteuerungsproblemen am Beispiel des Dynamischen Segelflugs. In: Informatik aktuell "Supercomputer '93", Springer-Verlag, Berlin, pp. 25-35, 1993.

[3] Sachs, G., Möller, H.: Visualization of the Separation Maneuver of Hypersonic Vehicles. Sonderforschungsbe-reich 255 "Transatmosphärische Flugsysteme" der DFG, Technische Universität München, Space Course 1993 Proceedings, pp. D2-1 - D2-5, 1993.

[4] Bulirsch, R.: Die Mehrzielmethode zur numerischen Lösung von Randwertproblemen und Aufgaben der optimalen Steuerung. Bericht der Carl-Cranz-Gesellschaft, Oberpfaffenhofen, 1971.

[5] Bock, H.G.: Numerische Behandlung von zustandsbeschränkten und Chebychef-Steuerungsproblemen. Kurs R 1.06 der Carl-Cranz-Gesellschaft, Oberpfaffenhofen, 1983.

[6] Oberle, H.J.: Numerische Berechnung optimaler Steuerungen von Heizung und Kühlung für ein realistisches Sonnenhausmodell, Institut für Mathematik der Technischen Universität München, TUM-M8310, 1983.

[7] Schoder, W. and Rochholz, H.: 1993, "Separation of Two-Stage Hypersonic Vehicles", Technische Universität München, Sonderforschungsbereich 255 "Transatmosphärische Flugsysteme", Space Course 1993 Proceedings, pp. 36-1 - 36-21.

[8] Sachs, G. and Schoder, W: Robust Control of the Separation of Hypersonic Lifting Vehicles. AIAA Paper No. 92-5013, AIAA 4th International Aerospace Planes Conference, Orlando, Florida, Dec. 1-4, 1992.

[9] Rayleigh, J.W.S.: The soaring of birds. Nature 27, pp. 534-535, 1883.

[10] Idrac, P: Ètude théoretique des manéurves des albatros par vent croissant avec L'altitude. C.r. hebd. Sanc. Acad. Scr., Paris 179, pp. 1136-1139, 1924.

[11] Prandtl, L.: Beobachtungen über den dynamischen Segelflug. Zeitschrift für Flugtechnik und Motorluftschiffahrt, Bd., 21, p. 116, 1930.

[12] Pennycuick, C.J.: The flight of petrels and albatrosses (Procellarinformes), observed in South Georgia and its vicinity. Phil. Trans. R. Soc. Lond. B 300, pp. 75-106, 1982.

[13] Nottebaum, T. Goebel, O.: Simulation optimaler Flugbahnen des dynamischen Segelflugs und Auslegung eines Modellflugzeugs. Zeitschrift für Flugwissenschaften und Weltraumforschung 13, S. 48-56, 1989.

[14] Sachs G.: Minimalbedingungen für den dynamischen Segelflug. Zeitschrift für Flugwissenschaften und Weltraumforschung 13, pp. 188-198, 1989.

EFFICIENT METHODS AND PARALLEL COMPUTING IN NUMERICAL FLUID MECHANICS

M. Schäfer
Lehrstuhl für Strömungsmechanik
Universität Erlangen-Nürnberg
Cauerstr. 4, D-91058 Erlangen

SUMMARY

In this paper the application of modern techniques of high performance scientific computing to computational fluid dynamics is considered. After an introduction into the general concepts for the numerical simulation of laminar and turbulent flows, the reasons for the very high computational effort for such computations are discussed. Acceleration techniques like multi-grid methods, vectorization, and parallelization are presented and their performance is demonstrated. It is shown that the efficiency of the simulations can be improved by orders of magnitude. Considering two three-dimensional applications the necessity of accurate discretizations and the practical possibilities using advanced algorithms together with modern parallel computer architectures are illustrated.

INTRODUCTION

Flow processes are of major significance in many technical systems, since they often are crucial with respect to their efficiency and economy. The investigation and optimization of these processes by numerical simulation becomes more and more attractive, since the recent progresses in high performance scientific computing, which are due to both advances in numerical methods and advances in computer technology, offer new possibilities for complex fluid flow simulations. In comparison to experimental techniques (e.g. wind tunnels) with informations from fast and reliable simulations, together with a few reference experiments, a considerable reduction of expenses and development times can be achieved, a fact that is of great importance with respect to industrial competitiveness requiring fast reactions on market demands for new and better products or more efficient process technology. Important areas of applications in this context are for instance:

- aerodynamics of motor vehicles, planes, and spacecrafts,
- turbine engineering, combustion processes, and plant safety,
- heat exchangers, pumps, valves, nozzles and ventilation systems,

- deposition and etching processes in semiconductor technology,
- sewage and environment techniques, weather and climate forecast,
- microsystem technology, biomechanics, medicine, etc.

Due to this broad field of applications and its great industrial importance the numerical simulation of fluid flows plays a central role in the relatively new discipline of high performance scientific computing.

Because of complex physical and chemical phenomena (e.g. turbulence, combustion, radiation) and complex geometries (e.g. airplanes, engines, chemical reactors), fluid flow processes with heat and mass transfer usually are very complex, and their numerical treatment requires a very large amount of computational effort. Therefore, to deal successfully with such problems the most efficient numerical techniques together with the most powerful computer architectures have to be used.

In principle there are two possibilities to reduce the computational effort for any kind of numerical simulation:

- use of more efficient numerical algorithms,
- use of more powerful computer hardware.

In both areas great progress were made in the last ten years, which in summary has led to an increase in the performance of fluid flow simulations by orders of magnitude. Usually algorithmic and hardware aspects must be considered in close relation, since an algorithm running efficiently on one computer system must not necessarily be the best choice for an other one. It is the combination, i.e. that efficient algorithms run efficiently on powerful computer architectures, to which special attention has to be paid and which should be understood by high performance scientific computing.

In this paper we will discuss the most relevant aspects concerning the application of modern techniques of high performance scientific computing to computational fluid dynamics. We give an introduction into the general concepts of fluid flow simulations and show how the very high computational effort for such computations can be reduced by orders of magnitude employing acceleration techniques like multi-grid methods, vectorization, and parallelization. Finally, the necessity of accurate discretizations in order to obtain reliable and physically correct predictions, and the capabilities using advanced algorithms together with modern parallel computer architectures, are illustrated by considering two three-dimensional applications.

All numerical results presented in the paper are obtained with fluid dynamics codes developed at the Department of Fluid Mechanics of the University Erlangen-Nürnberg.

BASIC CONCEPTS FOR FLUID FLOW COMPUTATIONS

The detailed mathematical modelling of fluid flow processes, the base for their successful numerical simulation, results in a complex system of nonlinear partial differential equations with complex boundary conditions. The equations can be derived from the basic conservation laws of continuum mechanics for mass, momentum, energy, and species. For incompressible Newtonian fluids the equations take, for instance, the following form:

$$\frac{\partial}{\partial x_j}(\rho U_j) = 0, \tag{1}$$

$$\frac{\partial(\rho U_i)}{\partial t} + \frac{\partial}{\partial x_j}\left(\rho U_j U_i - \mu\left(\frac{\partial U_i}{\partial x_j} + \frac{\partial U_j}{\partial x_i}\right)\right) + \frac{\partial P}{\partial x_i} = \rho g_i \tag{2}$$

$$\frac{\partial(\rho T)}{\partial t} + \frac{\partial}{\partial x_j}\left(\rho U_j T - \frac{\mu}{Pr}\frac{\partial T}{\partial x_j}\right) = R_T, \tag{3}$$

$$\frac{\partial(\rho C_k)}{\partial t} + \frac{\partial}{\partial x_j}\left(\rho U_j C_k - \rho D_k \frac{\partial C_k}{\partial x_j}\right) = R_k. \tag{4}$$

The unknowns in these system are the velocity vector $U = (U_1, U_2, U_3)$, the pressure P, the temperature T, and concentrations of species C_k $(k = 1, \ldots, K)$. The dynamic viscosity μ, the density ρ, the source terms R_k and R_T, the Prandtl number Pr, the diffusion coefficients D_k, and the gravitational acceleration vector $g = (g_1, g_2, g_3)$ are problem dependent physical quantities which, in general, can also depend on the unknowns. The first two equations (balance of mass and momentum) are known as *Navier-Stokes equations.*

To solve the above equation system numerically, the problem region has to be covered by a numerical grid with respect to space and time. Using some kind of discretization process, either finite difference, finite element, or finite volume methods, the system is then approximated by a nonlinear system of algebraic equations.

For finite volume methods, which have the advantage to be fully conservative, the flow region is divided into a finite number of control volumes. The transport equations, which all are of the general form

$$\frac{\partial(\rho\Phi)}{\partial t} + \frac{\partial}{\partial x_j}\left(\rho U_j \Phi + \Gamma_\Phi \frac{\partial \Phi}{\partial x_j}\right) = S_\Phi, \tag{5}$$

are then integrated over each control volume V and transformed by means of the theorem of Gauß into an integral equation

$$\int_V \frac{\partial(\rho\Phi)}{\partial t} dV + \int_A \left(\rho U_n \Phi + \Gamma_\Phi \frac{\partial \Phi}{\partial n}\right) dA = \int_V S_\Phi dV, \tag{6}$$

where A denotes the surface of the control volume and n the normal vector to it. After discretizing the fluxes in these balance equations one obtains a nonlinear algebraic equation of the form

$$a_P(\Phi)\Phi_P = \sum_{nb} a_{nb}(\Phi)\Phi_{nb} + b(\Phi), \tag{7}$$

where the sum runs over the neighbouring control volumes. Considered over the whole flow domain, i.e. for each control volume, the nonlinear algebraic system for the values of the unknowns, e.g. in the centres of the control volumes, results. For unsteady problems such a system has to be solved for each time step.

The accuracy of the discretization, i.e. how good the algebraic system approximates the differential equation, is mainly determined by the number of grid points (control

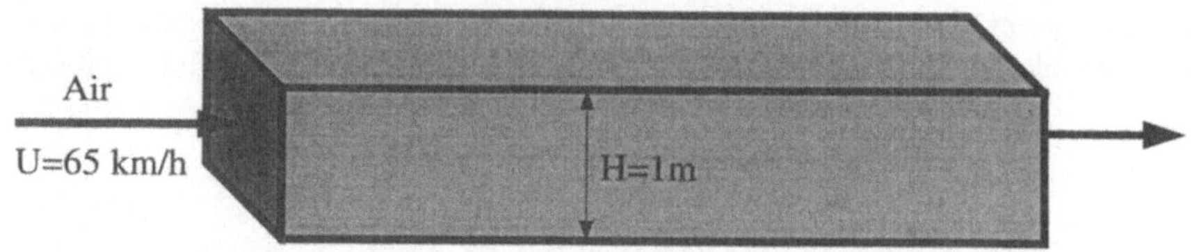

Figure 1 *Turbulent flow through a rectangular channel.*

volumes). The complexer the flow structure, i.e. the more and smaller the eddies and the larger the gradients of the flow quantities, the more difficult it is to resolve the flow domain with the fineness required for reaching a certain accuracy. It is well known that too coarse grids, too large times steps, or too short time intervals can yield physically incorrect results.

For turbulent flows, which dominate in practical applications, tremendous numbers of grid points and time steps are necessary to yield reliable solutions. To illustrate the effort, we consider the very simple situation of a turbulent flow of air through a rectangular channel shown in Fig. 1. ¿From theoretical investigations it is known that the size of the smallest eddies in this flow is about $0.2mm$. Assuming that the adequate resolution of these eddies requires 10 grid points in each space direction a number of 10^{14} grid points results. To obtain meaningful average values typically 10^4 time steps are necessary. Since usually about 500 floating point operations per grid point and time step are required to solve the equations, the total number of operations amounts to 10^{20}. Considering that the current most powerful computer systems have a performance of about 10 Billion floating point operations per second (10 GFlops) we end up at a total computing time of approximately 2000 years for solving our problem on such a machine.

An important quantity when discussing the computational effort for flow simulations is the Reynolds number defined by

$$Re = \frac{characteristic\ velocity * characteristic\ length}{density * dynamic\ viscosity} \tag{8}$$

indicating the relative magnitude of inertia forces and viscous forces. It can be shown that, in general, for the numerical solution of the problems the memory requirements increase with $Re^{2.6}$ and the computing time with $Re^{3.5}$. For the above channel flow we have $Re = 10^6$, only a medium sized value in applications.

The example shows that even for simple problems the direct simulation of turbulence is out of reach today. Therefore, in the case of practical turbulent flows, some kind of additional modelling of the turbulence must be incorporated into the mathematical problem description. Here, there currently exist mainly two approaches: the large eddy simulation and the turbulence models.

Since for complex problems also the large eddy simulation is beyond the capabilities of current supercomputers, the use of turbulence models dominates in current practice. To give an idea of such an approach we briefly outline the basic concept of the widely spread k-ε-model. The flow quantities and the equations are averaged with respect to time and the resulting extra stresses are modelled by relations involving the turbulent

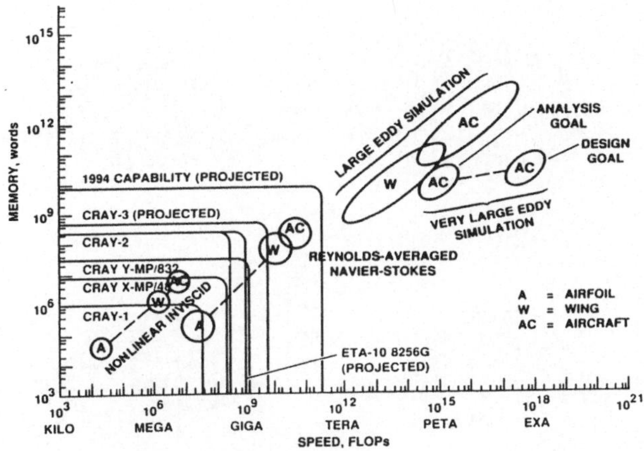

Figure 2 *Computer speed and memory requirements for flow simulation with different geometrical complexities and turbulence treatments (following [3]).*

kinetic energy k and the dissipation rate ε as additional unknowns:

$$\frac{\partial(\rho k)}{\partial t} + \frac{\partial}{\partial x_j}\left(\rho U_j k - \frac{\mu_t}{\sigma_k}\frac{\partial k}{\partial x_j}\right) = G - \rho\varepsilon, \tag{9}$$

$$\frac{\partial(\rho \varepsilon)}{\partial t} + \frac{\partial}{\partial x_j}\left(\rho U_j \varepsilon - \frac{\mu_t}{\sigma_\epsilon}\frac{\partial \varepsilon}{\partial x_j}\right) = \frac{\varepsilon}{k}\left(C_1 G - C_2\varepsilon\right) \tag{10}$$

with the production rate G and the eddy viscosity μ_t defined by

$$G = \mu_t\left(\frac{\partial U_i}{\partial x_j} + \frac{\partial U_j}{\partial x_i}\right)\frac{\partial U_i}{\partial x_j} \quad \text{and} \quad \mu_t = \rho C_\mu \frac{k^2}{\varepsilon}. \tag{11}$$

More advanced models like the so called Reynolds stress models require the solution of additional transport equations for further turbulence quantities.

In Fig. 2, an actualized version of a diagram in [3], the computational effort and the memory requirements are illustrated for the application of numerical simulation in aerodynamics. Considered are different geometrical complexities (airfoil, wing, aircraft) and different complexities of turbulence treatment. The values given are based on a computing time of 1 hour for one simulation run, which is a realistic value when the simulation is employed as a design tool usually requiring solutions for large numbers of parameter sets.

Due to the situation described above, possibilities for the acceleration of flow computations, in the past and nowadays, are intensively investigated in many research institutes. In the next two sections some of the approaches in this direction will be described.

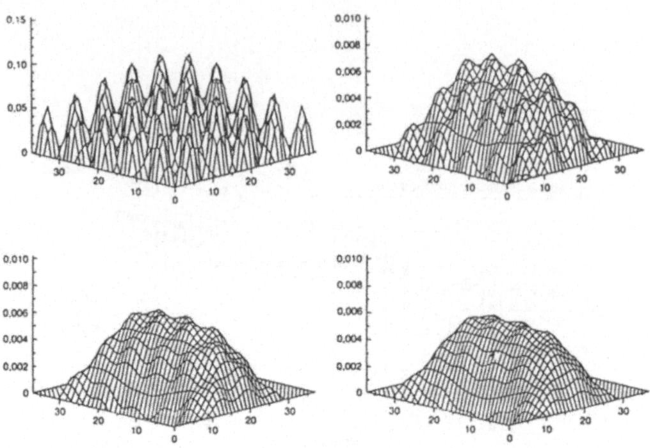

Figure 3 *Solution errors over a square domain after 4 consecutive iterations with a standard single-grid method (from [13]).*

ADVANCED ALGORITHMS

Concerning the improvement of numerical algorithms for flow computations a lot of approaches for the various stages of the solution process (e.g. grid generation, adaptive refinement, linear system solver, discretization scheme, linearization procedure, coupling of variables, etc.) could be mentioned. To illustrate the possibilities let us consider, as an important example, the so called multi-grid approach.

The coupled nonlinear algebraic equation system resulting from the discretization of the flow equations is usually solved by some kind of iterative procedure as for instance a pressure-correction method (e.g. [3]). If such an iterative scheme is applied to the system on a given grid, it turns out that only those frequencies of the solution error can be reduced efficiently, which correspond to the grid spacing. This behaviour is illustrated in Fig. 3 (taken from [13]), which shows the solution errors over a square domain after 4 consecutive iterations with a standard iterative scheme. The high frequencies (the peaks) are reduced after a few iterations, while the low frequencies nearly remain unchanged. It can be shown that due to this behaviour the computational effort to solve the system with such a method is proportional to N^2, where N is the number of grid points.

The basic idea of a multi-grid method is to involve a hierarchy of successively coarsened grids (see Fig. 4) into the iterative solution process. Following an adequate strategy for the movement through the different grid levels, for instance V-cycles or W-cycles (e.g. [6]), and transfering data consistently with the discretization scheme between the grids, this results in an efficient error reduction over a wide spectrum of frequencies. An iterative scheme then acts as a smoother of the solution error on the different grid levels. The effect of such a multi-grid approach is that the convergence rate becomes independent on the grid spacing and that the computational effort only increases linearly with N.

Figure 4 *Hierarchy of successively coarsened grids for a multi-grid method.*

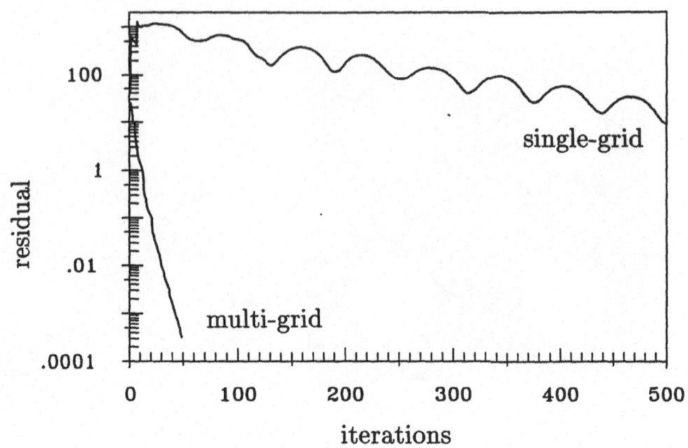

Figure 5 *Comparison of typical convergence histories of single-grid and multi-grid method.*

Table 1 *Typical acceleration factors with a multi-grid method for* 10^5 *grid points.*

	2-d (5 grids)	3-d (3 grids)
stationary flow	80-120	40-60
instationary flow	20-40	10-20

In Fig. 5 the convergence history of a single-grid method and a multi-grid method is compared for a typical flow computation. One can see the enormous improvement of the convergence rate by the multi-grid technique, resulting in a significant acceleration of the computation. The finer the grid, the larger becomes the improvement. In Tab. 1 typical acceleration factors for a grid with 10^5 points are indicated.

Multi-grid methods have now proven to work efficiently for a variety of practical applications (e.g. [1],[8],[11]) and they must be considered as the most relevant algorithmic development with respect to improvement of fluid flow computations in the recent years.

ADVANCED COMPUTER ARCHITECTURES

Let us now turn to improvements of numerical simulations due to more efficient computer hardware. Besides the improvements in the clock rates of the processors, two key words must be named in this context: *vectorization* and *parallelization*. Before going into more detail with these techniques, one general comment should be addressed: the major objective using these techniques should be to implement efficient algorithms running efficiently on vector and parallel computers and not to use much less efficient but efficiently vectoriable and parallelizable algorithms. The effect, when this principle is not fulfilled will be illustrated later.

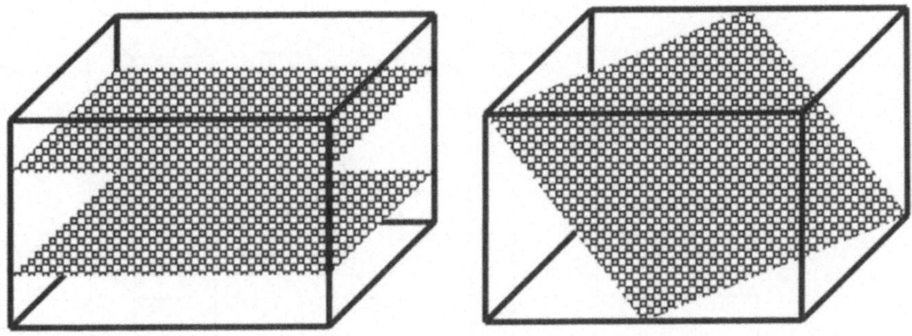

Figure 6 *Passing through a flow domain using a scalar (left) and vectorized (right) ILU algorithm.*

Table 2 *MFlop rates for a scalar and a vectorized 3-d flow simulation and theoretical peak performance for a Cray YMP and a SNI S200 (each for one processor).*

Computer	Scalar code	Vector code	Peak
Cray YMP	14	181	333
SNI S200	15	322	960

We first take a look to the vectorization. While vector machines are established now since several years, the efficient implementation of algorithms on such machines still requires special attention and is a topic of actual research. The main problem when vectorizing an algorithm is to avoid recursiveness and data dependencies in order to obtain large vector lengths for an efficient utilization of the vector pipelines.

As an example we consider ILU algorithms (e.g. [6]) for the relaxation of linear systems of equations, which usually is the most time consuming part of a flow computation. These algorithms are strongly recursive (a fact that is mainly responsible for their high efficiency), in the way that before a point can be calculated some of the neighbouring points must already be computed. The recursion can partly be resolved by marching by diagonal planes through the solution domain (see Fig. 6). This gives a vector length equal to the number of points in these planes, which already for medium sized problems (i.e. $50 \times 50 \times 50$ grid points) is large enough to work in the saturation region of the vector pipelines of current supercomputers.

In order to give an example for the possible acceleration by vectorization in Tab. 2 the MFlop rates for a typical 3-d computation (50000 grid points) are indicated for a scalar and a vectorized version of a finite volume multi-grid code on two modern vector supercomputers. For reference the theoretical peak performances of the machines are also given. It should be noted that in this case the vectorized code runs with the same numerical efficiency as the scalar one (with an ILU solver and a multi-grid technique). It can be seen that an acceleration of more than one order of magnitude can be achieved and that it is possible to perform an efficient flow computation quite close to the theoretical peak performance of vector machines.

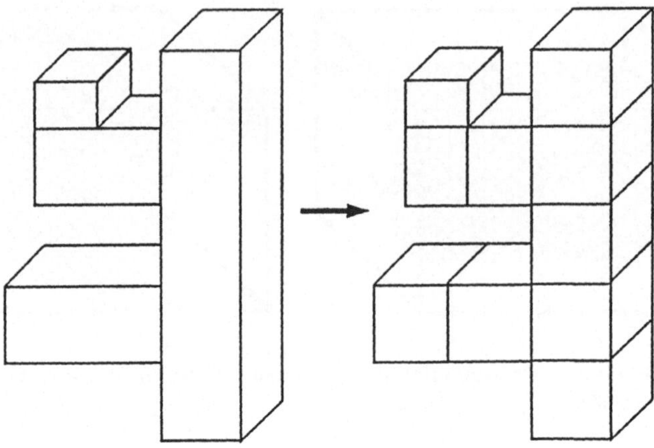

Figure 7 *Strategy for parallelizing flow computations in complex geometries using blockstructured grids.*

Since there are physical limits of the performance of a single processor, which are nearly reached by modern supercomputers, the possibilities of using more than one processor working in parallel on a flow problem has become a topic of intensive research in the recent years and it has turned out that, if algorithms are specially designed for this purpose, there lies an enormous potential in such an approach.

Concerning the parallelization of fluid flow computations mainly the following techniques are currently investigated: grid partitioning (e.g. [2]), domain decomposition (e.g. [5]), combination techniques (e.g. [4]), and time parallel methods (e.g. [10]). We will restrict ourselves here to a brief description of the grid partitioning technique, which presently is the most popular approach.

In the context of grid partitioning the concept of blockstructured grids, which is closely related, has to be mentioned. Blockstructured grids can be viewed as a compromise between the flexibility of fully unstructured grids and the numerical efficiency of globally structured grids, and, besides the possibility of modelling complex geometries occuring in practical applications, this technique also provides a natural basis for the parallelization of flow computations by means of grid partitioning. A possible strategy, which has been applied successfully to various flow problems (e.g. [2],[7],[9]), is illustrated in Fig. 7. The blockstructure required to model the geometry is mapped by an automatic load balancing procedure into a parallel blockstructure such that the resulting subdomains can be assigned to individual processors. To ensure the coupling of the subdomains boundary data of neighbouring blocks are required. In order to keep the communication effort for this as low as possible, along the block interfaces auxiliary control volumes containing the corresponding boundary values of the neighbouring blocks are introduced. This allows for carrying out most of the computations locally within the individual blocks (i.e. processors) and the boundary values have to be transfered only from time to time during the iterative algorithm (see [12]).

A measure for the performance of an implementation of a parallel algorithm is the total efficiency defined by

$$E_{tot}^{n} = \frac{T_s}{nT_p^n}, \tag{12}$$

where n is the number of processors, T_s is the computing time of the best serial algorithm, and T_p^n is the computing time of the parallel algorithm on n processors. Important influence factors for this efficiency are the possible decrease of the rate of convergence due to modifications in the algorithm required for parallelization (numerical efficiency), the additional effort for the interprocessor communication (parallel efficiency), and the different utilization of processors (load balancing efficiency). Regarding these aspects, and also the very rapidly developping parallel computer market, the following principles are of major importance when designing a parallel algorithm:

- retaining the high efficiency of serial methods,
- minimization of the communication effort,
- balanced utilization of the processors,
- portability and scalability of the code.

Usnd NX/2 communicationjjjually it is not possible to fulfil all of these requirements optimally at the same time, and the main task is to find a compromise yielding the best performance of a specific algorithm on a wide range of parallel machines.

In order to show the capabilities of current parallel computers with repect to the acceleration of flow computations and also to illustrate the influence of different hardware and software properties of such machines on the performance of the computations, we present results for a fluid flow benchmark problem on different parallel architectures.

The considered problem is a steady natural convection flow in an inclined cavity (e.g. [12]). For discretization a non-orthogonal non-equidistant grid with 320×320 control volumes is employed and the problem is solved with a full multi-grid method with 6 grid levels (10×10 on the coarsest grid) and pressure-correction smoothing. The following computer systems are included in the comparison:

- a cluster of SUN Sparc 10/20 workstations with Ethernet connection and TCGMSG communication software,
- a Parsytec SuperCluster with T805 transputers and Parix communication software,
- a virtual shared memory Siemens-Nixdorf KSR1 with Pthreads communication,
- an Intel iPSC with i860 vector nodes and NX/2 communication software.

It should be remarked that the code runs on these quite different computer architectures with minor modifications, only specific communication primitives have to be adapted. In Tab. 3 the total computing times for solving the problem and the corresponding efficiencies are given for the various machines with different numbers of processors. As reference values the computing time on a single SUN Sparc 10/20 is also indicated.

There are three major factors that are responsible for the performance of the parallel machines for the considered kind of flow computations: the sustained Flop rate of the

Table 3 *Total computing times and efficiencies for a benchmark problem on various hardware platforms with different numbers of processors.*

Computer	Proc.	CPU time (s)	Efficiency (%)
SUN Sparc 10/20	1	589	960
SUN Sparc 10/20	5	243	49
SNI KSR1	10	88	85
Parsytec SC	25	278	90
Intel iPSC	25	73	76
Parsytec SC	100	120	52

Table 4 *Computing times and efficiencies for a benchmark problem with a single-grid and a multi-grid method.*

Method	CPU time (s)	Efficiency (%)
single-grid	103000	96
multi-grid	120	49

node processors, the latency time for a communication, and the number of processors. The influences of these parameters are closely related to each other and, therefore, must be taken into account simultaneously, together with algorithmic aspects, when evaluating the performance of a parallel machine with respect to flow computations. For instance, the larger the number of processors and the communication requirements of the algorithm the higher is the importance of a small latency time. Of course, in general, the performance of the parallel computer increase with higher Flop rates and lower latency times, but not necessarily with a larger processor number. Depending on the ratio of the Flop rate and the latency time the number of processors which can be used for accelerating the computation of a specific problem is limited.

It should be noted that the efficiency defined above is an adequate measure for the evaluation of the performance of the parallel implementation of an algorithm, but that for the user only the computing time is relevant. To make this more concrete, in Tab. 4 the computing times and the efficiencies for the computation of the above benchmark problem on a Parsytec SC with 100 processors are indicated for a single-grid and a multi-grid method. Since on the coarser grids the portion of the communication time compared to the arithmetic work is higher, the efficiency is much lower for the multi-grid case, but if we look at the computing times, it can be seen that due to its much higher numerical efficiency the multi-grid method comes out much more efficient for the user.

APPLICATIONS

To illustrate the capabilities of the approaches considered in the previous sections we consider two examples from the very broad field of applications for numerical simulation

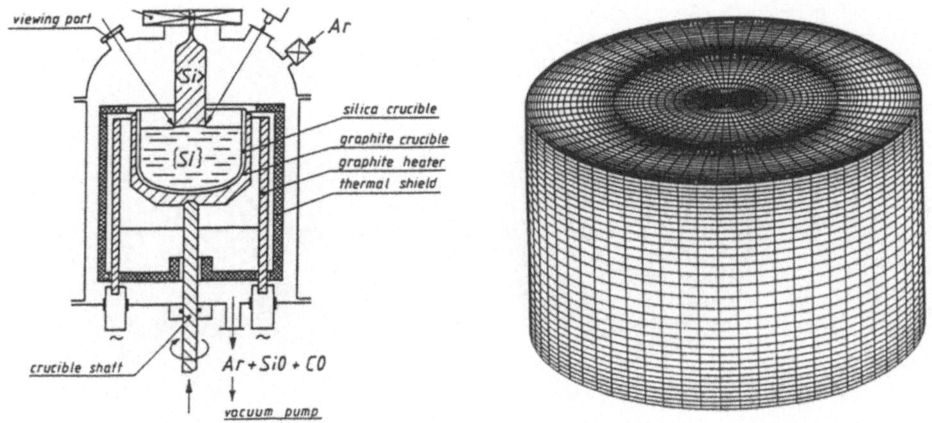

Figure 8 *Czochralski crystal growth configuration and numerical grid used for the modelling of the melt flow.*

of fluid flows. The first problem concerns the simulation of the melt flow in a Czochralski crystal growth configuration for the manufacturing of semiconductor crystals (see the article of G. Müller in this volume for a details of this process). Figure 8 shows a sketch of the configuration and the numerical grid used for the discretization of the melt.

We simulate the flow with a 3-level time-dependent multi-grid scheme with about 80000 control volumes on the finest grid. Figure 9 shows the velocity field directly below the crystal after a real time of 210s and 370s. After approximately 250s the flow becomes unsteady and four eddies precessing around the symmetry axis of the crucible appear. The periodic behaviour was also found from experimental investigations. It should be noted that on the two coarse grids a steady axisymmetric flow is obtained. Thus, we have here an example that a simulation produce physically incorrect results, if either the grid is too coarse or the considered time interval in which the process is simulated is too short.

The acceleration of the simulations, which can be achieved by the techniques discussed in the previous section, are indicated in Tab. 5, where the computing times for a simulation of 8 minutes real time are given. The time for the parallel simulation is based on 100 Cray processors running with an efficiency of 56 percent (a realistic value cf. Tab. 3). Compared to a scalar serial single-grid computation the vectorized parallel multi-grid computation is about 7500 times faster.

As a second example we consider the buoyancy driven flow in a rectangular cavity with a heated obstacle, a flow situation which occurs in a variety of applications. In Fig. 10 the problem geometry is shown together with a sketch of the occuring complex flow pattern. Again, if the numerical grid is too coarse the simulation does not yield a physically correct flow pattern. This is illustrated in Fig. 11, which shows the velocity vectors in a vertical cross section computed on a grid with 13440 (top) and with 107520 (bottom) control volumes. On the coarser grid one obtains 6 pairs of convection cells, while on the finer grid 7 pairs occur, which is in agreement with experimental results.

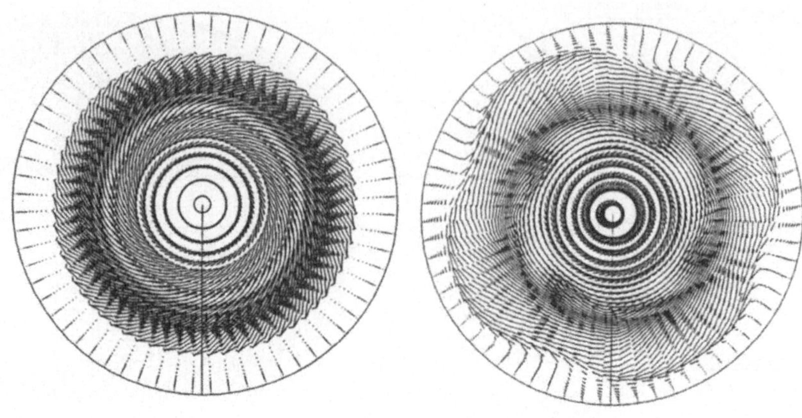

Figure 9 *Velocity field directly below the crystal after a real time of 210s (left) and 370s (right).*

Table 5 *Acceleration of numerical simulation of a Czochralski melt flow by multi-grid methods, vectorization, and parallelization.*

Method	CPU time
single-grid, scalar, serial	$\approx$ 3 weeks
+ multi-grid	$\approx$ 2 days
+ vectorization	4 hours
+ parallelization (100 proc.)	$\approx$ 4 minutes

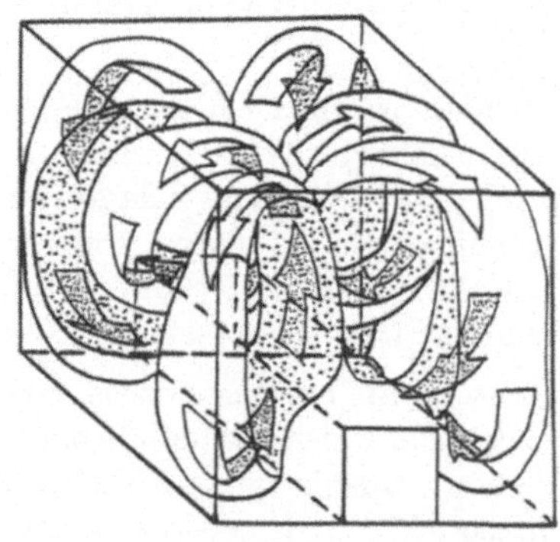

Figure 10 *Buoyancy driven flow in a rectangular cavity with a heated obstacle.*

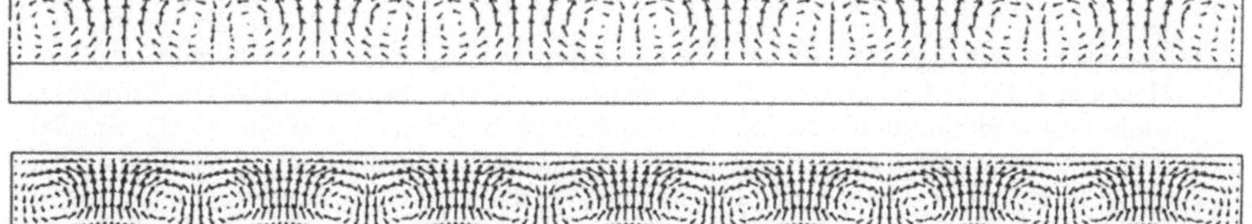

Figure 11 *Predicted flow pattern for the buoyancy driven cavity flow on a coarse grid (top) and a fine grid (bottom).*

Table 6 *Acceleration of numerical simulation of a buoyancy driven cavity flow by multi-grid methods, vectorization, and parallelization.*

Method	CPU time
single-grid, scalar, serial	$\approx$ 4 months
+ multi-grid	$\approx$ 2 days
+ vectorization	3 hours
+ parallelization (100 proc.)	$\approx$ 3 minutes

For this problem the improvement by the acceleration techniques is even more significant as for the Czochralski problem. The corresponding computing times are indicated in Tab. 6 showing that in summary an acceleration of about a factor 57000 is achieved.

CONCLUSIONS

It has been shown that fluid flow simulations, for which a broad field of applications exist, are very ambitious with respect to computer resources. With modern techniques of high performance scientific computing like multi-grid methods, vectorization, and parallelization a drastical accleration of the computations can be achieved, which opens new perspectives for the application of numerical simulation to the investigation and optimization of industrial processes involving fluid flows. The simulations become very cost and time effective and usually they are more universally applicable as pure experimental investigations. Due to these aspects and with regard to industrial competitiveness, there is no doubt that simulation techniques, accompanied by a few reference experiments, will play a more and more important role in industry.

ACKNOWLEDGMENTS

The author would like to thank all members of the research group "Numerical Fluid Mechanics" of the Department of Fluid Mechanics of the University Erlangen-Nürnberg, who developed the numerical methods and software referenced in this work. Special thanks are addressed to E. Schreck for carrying out the computations on the parallel computers, to H.J. Leister for providing the results for the applications, as well as to F. Durst for many helpful discussions and his untiring efforts to establish the *Bavarian Consortium of High Performance Scientific Computing (FORTWIHR)*, which with its financial support provides the framework for this work.

REFERENCES

[1] F. Durst, L. Kadinskii, M. Perić, and M. Schäfer. Numerical Study of Transport Phenomena in MOCVD Reactors Using a Finite Volume Multigrid Solver. *Journal of Crystal Growth*, 1993. 612-626.

[2] F. Durst, M. Perić, M. Schäfer, and E. Schreck. Parallelization of Efficient Numerical Methods for Flows in Complex Geometries. In *Flow Simulation with High-Performance Computers I*, Notes on Numerical Fluid Mechanics, pages 79–92. Vieweg Verlag, 1993.

[3] C.A.J. Fletcher. *Computational Techniques for Fluid Dynamics.* Springer, Berlin, 1988.

[4] M. Griebel. Sparse Grid multilevel methods, their parallelization, and their application to CFD. In R. Pelz, A. Ecer, and J. Häuser, editors, *Parallel Computational Fluid Mechanics '92*, pages 161–174. North-Holland, 1993.

[5] W. Gropp and D. Keyes. Domain decomposition methods in computational fluid dynamics. *Int. J. Num. Meth. in Fluids*, 14:147–165, 1992.

[6] W. Hackbusch. *Multi-Grid Methods and Applications.* Springer, Berlin, 1985.

[7] M. Hortmann, I. Janzik, L. Kadinski, M. Schäfer, and H. Scheidat. A Parallel Multigrid Algorithm for Flow Computations with Thermal Solid/Fluid Interactions. In C. Taylor, editor, *Numerical Methods in Laminar and Turbulent Flow VIII*, pages 1459–1470, Swansea, 1993. Pineridge Press.

[8] M. Hortmann, M. Perić, and G. Scheuerer. Finite volume multigrid prediction of laminar natural convection: Benchmark solutions. *Int. J. Num. Meth. in Fluids*, 11:189–207, 1990.

[9] M. Hortmann and M. Schäfer. Numerical prediction of laminar flow in plane, bifurcating channels. *Computational Fluid Mechanics*, 1994. To appear.

[10] G. Horton. TIPSI - a time-parallel SIMPLE-based method for the incopressible Navier-Stokes equations. In *Parallel Computational Fluid Dynamics 91*, pages 243–256, Amsterdam, 1992. Elsevier.

[11] H.J. Leister and M. Perić. Numerical Simulation of a 3D Czochralski-Melt Flow by a Finite Volume Multigrid-Algorithm. *Journal of Crystal Growth*, 123:567–574, 1992.

[12] M. Perić, M. Schäfer, and E. Schreck. Numerical simulation of complex fluid flows on MIMD computers. In *Parallel Computational Fluid Dynamics 92*, pages 311–324, Amsterdam, 1993. Elsevier.

[13] G. Wittum. Mehrgitterverfahren. *Spektrum der Wissenschaft*, 4, 1990.

Addresses of the Editors of the Series "Notes on Numerical Fluid Mechanics"

Prof. Dr. Ernst Heinrich Hirschel (General Editor)
Herzog-Heinrich-Weg 6
D-85604 Zorneding
Federal Republic of Germany

Prof. Dr. Kozo Fujii
High-Speed Aerodynamics Div.
The ISAS
Yoshinodai 3-1-1, Sagamihara
Kanagawa 229
Japan

Prof. Dr. Bram van Leer
Department of Aerospace Engineering
The University of Michigan
Ann Arbor, MI 48109-2140
USA

Prof. Dr. Keith William Morton
Oxford University Computing Laboratory
Numerical Analysis Group
8-11 Keble Road
Oxford OX1 3QD
Great Britain

Prof. Dr. Maurizio Pandolfi
Dipartimento di Ingegneria Aeronautica e Spaziale
Politecnico di Torino
Corso Duca Degli Abruzzi, 24
I-10129 Torino
Italy

Prof. Dr. Arthur Rizzi
Royal Institute of Technology
Aeronautical Engineering
Dept. of Vehicle Engineering
S-10044 Stockholm
Sweden

Dr. Bernard Roux
Institut de Mécanique des Fluides
Laboratoire Associé au C.R.N.S. LA 03
1, Rue Honnorat
F-13003 Marseille
France

Brief Instruction for Authors

Manuscripts should have well over 100 pages. As they will be reproduced photomechanically they should be typed with utmost care on special stationary which will be supplied on request.
In print, the size will be reduced linearly to approximately 75 per cent. Figures and diagrams should be lettered accordingly so as to produce letters not smaller than 2 mm in print. The same is valid for handwritten formulae. Manuscripts (in English) or proposals should be sent to the general editor, Prof. Dr. E. H. Hirschel, Herzog-Heinrich-Weg 6, D-85604 Zorneding.